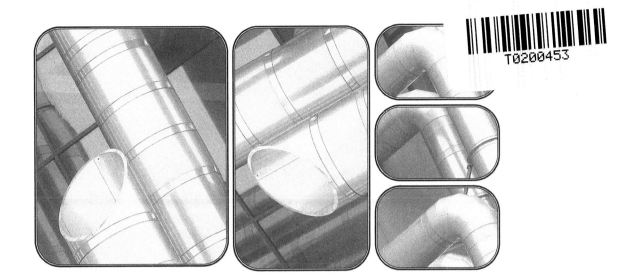

QUALITY CONCEPTS
FOR THE
PROCESS INDUSTRY

QUALITY CONCEPTS
FOR THE
PROCESS INDUSTRY

SECOND EDITION

Michael Speegle

DELMAR
CENGAGE Learning™

Australia • Brazil • Japan • Korea • Mexico • Singapore • Spain • United Kingdom • United States

Quality Concepts for the Process Industry, Second Edition
Michael Speegle

Vice President, Editorial: Dave Garza

Director of Learning Solutions:
Sandy Clark

Executive Editor: David Boelio

Managing Editor: Larry Main

Senior Product Manager:
Sharon Chambliss

Editorial Assistant: Lauren Stone

Vice President, Marketing:
Jennifer McAvey

Executive Marketing Manger:
Deborah S. Yarnell

Marketing Manager: Jimmy Stephens

Marketing Specialist: Mark Pierro

Production Director: Wendy Troeger

Production Manager: Mark Bernard

Content Project Manager:
Barbara LeFleur

Art Director: Benj Gleeksman

Technology Project Manager:
Chrstopher Catalina

Production Technology Analyst:
Thomas Stover

For product information and technology assistance, contact us at
Professional & Career Group Customer Support, 1-800-648-7450

For permission to use material from this text or product, submit all requests online at **www.cengage.com/permissions**.
Further permissions questions can be e-mailed to
permissionrequest@cengage.com.

Library of Congress Control Number: 2008937972

ISBN-13: 978-1435482449

ISBN-10: 1435482441

Delmar
5 Maxwell Drive
Clifton Park, NY 12065-2919
USA

Cengage Learning products are represented in Canada by Nelson Education, Ltd.

For your lifelong learning solutions, visit **delmar.cengage.com**

Visit our corporate website at **cengage.com.**

Notice to the Reader

Printed in the United States of America
3 4 5 6 7   26 25 24 23 22

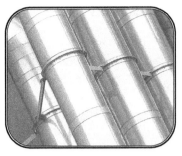

Table of Contents

Contents

About the Author

Michael Speegle has worked 18 years in the petrochemical and refining industry. He graduated from the University of Texas at Arlington with a Master of Arts degree in biology and a minor in chemistry. In the industry he has worked as a technician, supervisor, laboratory quality control technician, and training coordinator. He has worked two years as a technical writer for the process industry and is the author of several process technology textbooks, a novel, and several short stories and articles. He has had four industry internships and traveled to Alaska and offshore to study the exploration and production of oil and gas. Currently he is the department chairman for Process Technology, Instrumentation, and Electrical at San Jacinto College, central campus, in Pasadena, Texas.

For six years Mr. Speegle was a quality instructor for Amoco Chemical Company in Texas City, Texas. He became the lead instructor of a group of four instructors that taught all departments and divisions in the plant.

  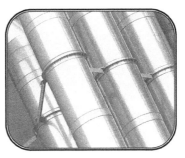

Preface

This course, and the accompanying book, has been designed and written specifically for the process industries, industries such as refining, petrochemicals, electric power generation, food processing and canning, and paper mills. Millions of dollars have been invested, and continue to be invested, in process industries. The investment is made with a belief and commitment to a return on that investment, in other words, a profit. If a company is producing the right product at the right time with little competition, profits come easily. However, when a company has a lot of competition, as do refineries and chemical plants, plus strict health, safety, and environmental constraints, profits do not come easily. Beginning in the late seventies many American companies sought a competitive advantage through embracing *quality*. Eventually, quality became the cornerstone of their competitive strategy.

Quality was desired and demanded of the manufactured product. Then company executives began to realize that to produce a finished product of high quality and *make a profit* required that the manufacturing process also be of high quality. Waste and repetitive mistakes had to be eliminated from the process. Better ways of doing a task had to be actively sought. Process technicians, with their intimate knowledge of the manufacturing process, could contribute substantially to the bottom line of a company. But that would not be possible without an understanding of what is meant by *quality,* how quality is achieved through continuous improvements, and several statistical quality tools. This course is a general overview of the history of quality, quality improvement processes and tools for improving quality. It was designed specifically for process technicians with the intent of making them a significant contributor to their company's competitive edge.

The reader will notice that many of the examples used and the group activities do not refer to refining or petrochemical scenarios. This is deliberate. The majority of students will have little knowledge of plant equipment and processes and be unable to relate to or contribute to the examples and activities. Because of this, familiar examples and scenarios were chosen to encourage learning and involvement.

Finally, I would like to acknowledge the help and advice of several kind friends, especially G.C. Shah, Mike King of Lubrizol, Evette Torres and Billy Collins of Shell Oil, and Cyndi Gillies of British Petroleum.

Mike Speegle
July, 2008

CHAPTER 1

Why Quality Is Important

Learning Objectives

After completing this chapter, you should be able to:

- *Explain how quality is used as a competitive tool.*

- *Explain why the system concept is necessary to quality systems.*

- *Explain why quality is, and always will be, a challenge to businesses.*

- *Explain why quality is compared to a never-ending marathon.*

- *State the advantage of having a quality program.*

INTRODUCTION

The word *quality* is everywhere. Television commercials emphasize the quality of a product or service of a company; magazines have glossy pages showing a husband and wife smiling with happiness at the quality of their new carpet or kitchen appliance. Someplace in a clinic or hospital, a statement reassures the public that the clinic or hospital has quality staff, quality service, and quality equipment. Did you ever consider why all the hoopla about quality? Why does every company supplying a product or service want the public to know it has a *quality* product or service? Isn't it enough that the company sells a product or service at a competitive price? This short chapter will answer these questions.

Bear in mind that the quality we are discussing is not that presented in frivolous ads that claim, "If you use our *quality* toothpaste, your teeth will be blinding white in two weeks," nor claims that, "If you use our quality skin lotion, your skin will look years younger in just weeks!" These are just examples of advertising and media hype, and they have no relationship with the quality systems required for a company to be a competitor in today's fierce business climate.

QUALITY AS A COMPETITIVE TOOL

Around the world, companies are intensively involved with quality because quality is a **competitive tool.** In other words, quality helps a company to compete better against other companies. Quality is a tool that allows a company to:

● Build, shape, and innovate its product or service.
● Unleash the talent of its workforce.
● Reduce waste.
● Increase efficiency and profits.

By doing any one or several of these actions, a company can be more competitive, be better able to survive in a fiercely competitive world, and be more likely to gain market share and increased profits.

The Law of the Jungle

An African fable illustrates the fierce competitive nature of business in today's world. In the fable, a lion wakes up hungry in the morning on the veldt. If the lion wishes to eat (to survive), he must be faster than the antelope that is his prey. That same morning, the antelope wakes up and knows that if he wants to survive another day he must be faster than the hungry lion. Both are in a race for survival, and the race continues every morning that the lion and the antelope awaken. This is the law of the jungle. Quality is critical to the business world because the business world is a jungle every bit as red in tooth and claw as a real jungle. However, instead of the bodies of slain animals, the business world jungle is littered with the bodies of bankrupt, failed, or downsized companies. They were "slain" because they were unable to compete well enough to survive, much less become dominant players in their fields. At the heart of their failure was a lack of quality in their business systems.

Some Automotive History Lessons

At one time in its history, the United States had more than twenty competing American automobile companies; now it has just three, and two of those (GM and Chrysler) are struggling to survive. I can remember one of my uncles complaining in 1972 that his daughter had bought a Honda. No one, the uncle exclaimed, had ever heard of a Honda. It looked like a green box on wheels! She would never get any resale value out of it! Talking to my cousin (the daughter), I found she was very pleased with her little car. Imagine; in 1972, the vast majority of Americans had never seen or heard of a Honda or Toyota automobile. Contrast that with what you see on the road today.

Note: Several times when I was in the military in Korea, I took taxis from the demilitarized zone to Seoul over rough gravel roads. I was very impressed with the way the little taxis took a licking and kept on ticking over roads that had ruts the size of bathtubs. I was so impressed that I consciously noted the names on the little cars. They were Nissan, Honda, and Toyota.

About twenty years ago, General Motors (GM) was the world's dominant car manufacturer, selling more cars and trucks than any other company in the world. The thought that another automobile manufacturer might surpass General Motors was inconceivable. Japanese automakers acquired 30 percent of the U.S. passenger car market in the 10 short years from the mid-seventies to the mid-eighties. In 2008, Toyota might become number one; it almost did in 2007. The year 2007 was a landmark for American automobile manufacturers. Ford Motor Company, the founder of the auto industry in the United States, lost its long-standing second place to Toyota, which sold more cars in the United States than Ford did. How did this happen? One automotive analyst said that from the 1970s to the 1990s, Detroit's attention was focused on finances, sales volume, debt repayment, and other factors—everything but the customers. So the customers voted with their wallets and began buying Japanese cars.

The American automotive industry today is an ongoing study of global competition and quality. GM is struggling to remain number one in automotive sales. Ford is struggling just to remain a dominant player in the automotive market. Chrysler was bought by Daimler in 1998, then was sold a few years later when Daimler realized it couldn't make the Chrysler unit profitable. Now Chrysler is privately owned. How did these three giants of American industry get in such a desperate fight for survival? Principally, they did not realize that competition was becoming global. They never anticipated that their customers, Americans, would choose foreign products over American-made products. They didn't realize that the customers are always looking for better deals, and most critically, they did not listen to their customers until it was almost too late. They relied on style and hype instead of reliability and quality. One reason many Americans are confirmed Toyota and Honda owners is because they are greatly reliable. **Table 1.1** reflects long-term reliability of four automobile manufacturers. Overall, American automobiles currently are improving in reliability and quality but have a ways to go before they catch up with Japanese automobiles. Most Americans once they buy a Toyota or Honda continue to buy Toyotas and Hondas.

There is a long list of American companies that no longer exist. Just a few of them are Compaq Computer, Montgomery Ward, Amoco, Nash, Eckerd, Arco, Rambler, Digital Equipment, and Wang. At one time, the executive officers and boards of directors of these companies felt certain their company was strongly positioned in the market and that failure was not conceivable. Today, all of these companies, like the dinosaurs, no longer exist.

Table 1.1 Cumulative Defects per 100 Vehicles

Vehicle Age	Toyota	Ford	GM	Chrysler
1	9	17	17	15
2	19	38	40	35
3	24	50	50	45
4	27	60	70	70
Note:				

QUALITY SYSTEMS

Today numerous quality systems are available for products and services. The very first non-military system was developed by Walter Shewhart, who attended the University of Illinois before receiving his doctorate in physics from the University of California at Berkeley in 1917. At that time, Bell Telephone's engineers had been working to improve the reliability of their transmission systems. Because amplifiers and other equipment had to be buried underground, a business was needed to reduce the frequency of failures and repairs. As well, tearing up streets and sidewalks to repair transmission systems was expensive.

When Dr. Shewhart joined the Western Electric Company Inspection Engineering Department at Hawthorne Works in 1918, industrial quality was limited to inspecting finished products and removing defective items. That all changed on May 16, 1924, when Dr. Shewhart prepared a page-long memorandum that described a simple diagram that today is recognized as a schematic of a control chart. That diagram, and the short text that preceded and followed it, set forth all of the essential principles and considerations that are involved in what we today know as ***process quality control.***

Shewhart's work pointed out the importance of reducing variation in a manufacturing process and understanding that continual process adjustment in reaction to nonconformance actually increased variation and degraded quality. Dr. Shewhart created the basis for the control chart and the concept of a state of statistical control, a concept used in most manufacturing processes today. The control chart identified acceptable and unacceptable variations that could harm quality. This was an early warning system to alert workers that they were about to produce bad product.

Since then, several quality systems have been developed. Some of the most familiar are W. Edwards Deming's, Joseph Juran's, Philip Crosby's, Kaoru Ishikawa's, William Conway Genichi Taguchi's, International Organization for Standardization (ISO), total quality management (TQM), American Society for Quality (ASQ), and Six Sigma. Businesses and corporations spend large sums to have their employees trained in some type of quality system (Table 1.2). A ***quality system*** is a methodology that causes or allows something desirable to happen or causes something undesirable to stop happening in a work process. A system is necessary because a problem (a lack of quality in some process) will not be solved unless it is approached systematically. Too often, the same mistakes occur over and over again, and

Table 1.2   Quality Systems

Military standards
Total quality management (TQM)
ISO 9000
Malcolm Baldrige Award
Deming Prize
Six Sigma and lean production
Note: Some people might include Juran, Crosby, and Deming and their tools, metrics, and philosophy as quality systems.

managers and workers become resigned to living with the problem rather than eliminating the problem. They don't have a quality improvement process; rather, they accept a certain consistent amount of failure as normal to a work process. As long as we are willing to accept errors, we will have errors. It is a self-fulfilling prophecy.

THE QUALITY MARATHON

It is erroneous to think that quality must be addressed with just a company's final product. Quality needs to be maintained in every aspect of a company, from its sales force, data processing, management, manufacturing site, raw materials suppliers, raw materials receiving and storage, and finished product shipment. If you don't believe that, remove any one of these processes and ask yourself what will be the effect on the company. A company can lose money and customers through any one of its many departments or divisions.

Most of today's baby boomers—people born between the years of 1946 and 1964—are astounded by the amount and rapidity of change since they were in high school. There is no comparison between today's world and the world of the 1950s. Two things caused this change—competition and innovation. China, India, Korea, Singapore, and Taiwan were not competing in manufacturing products or supplying services at that time. They are now all fierce competitors that supply products and services of very high quality. To sell their products, they must have customers—lots of customers, and unfortunately, they just might be your customers or mine. Also, unfortunately, only a limited supply of customers has adequate credit and money.

Today, every nation that manufactures items for sale on the world market is engaged in a never-ending marathon to win or at least place in the world market. Picture thousands of people running in the New York or Boston Marathon, sweating, ignoring pain and exhaustion, and determined to make it to the finish line. Now, instead of people in that race, substitute Boeing and Airbus, Toyota, Ford, Honda, General Motors, General Electric, Siemens, Sanyo, Sony, Oracle, SAP, and so on (Figure 1.1). They are all in a marathon that has no finish line, and they cannot afford to rest or they will be passed up. (Remember the fable of the tortoise and the hare?) If they fall too far behind, they will become consigned to the graveyard of companies that used to exist but are now found only in history books. A critical component that keeps them in this grueling marathon is quality. Quality is like the person standing on the sideline holding out a cup of Gatorade to the marathon runner. That cup fuels runners and keeps them in the race.

Once a company has established a quality system(s), they have just begun their marathon. The system(s) will constantly need to be improved and varied to meet the changes and challenges that confront that industry. An American company is competing not only against American companies but also against companies from Japan, Germany, South Korea, China, India, and so on. When your competitor makes a product similar to yours and sells it for about the same price, you might feel complacent and think everything is okay. At that time, you are running shoulder to shoulder in the marathon. However, when your competitor sells its product for 5 percent less than yours the next month and it is not on sale, you now have a serious problem. Your competitor is now several yards ahead and pulling away from you in the race. You are going to have to assess and tweak your systems (continuous improvement) so that you can afford to sell your product at the same price or cheaper. Now you are running shoulder to shoulder again, but you won't be able to relax because behind

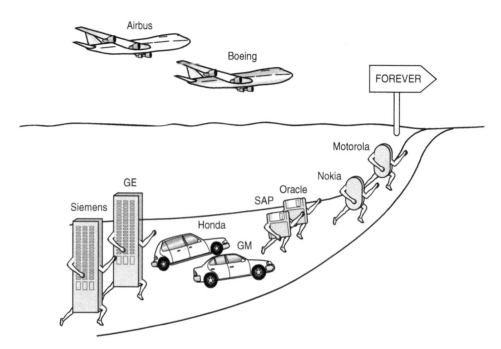

Figure 1.1   Manufacturing Marathon

you is the sound of gaining footsteps. Another competitor from the Far East is moving up in the race, seeking to gain your market share (customers).

This scenario is not far-fetched. When I was in college in the 1960s, I had never heard of any company from India that was even a moderate competitor with any American company. In January 2008, Tata Motors, an Indian company, unveiled the "Nano," a small passenger car that will be able to meet European pollution and automobile safety standards. It will average fifty miles per gallon of gasoline and sell for about $2,500, exclusive of shipping and taxes. The footsteps of Tata Motors can be heard coming up behind GM, Toyota, Honda, Mercedes, Ford, and Nissan. These companies didn't even know Tata Motors was in the race! Plus, in March 2008, Tata Motors bought Ford's Jaguar and Land Rover automotive divisions.

Don't laugh and wonder who would buy a $2,500 automobile called the Nano. It is well made, reliable, and pleasing to the eye—and it gets fifty miles to the gallon. Millions of people would like a $2,500 new car, and they could pay cash for it. You can bet GM, Toyota, Honda, and Mercedes will have their engineers study the car thoroughly to assess the probability of its becoming a winner and to avoid possible loss of market share.

DEFINING QUALITY

Defining what *quality* is can be difficult. *Quantity* is easy to define because it is a finite number. If you say six, everyone who understands your language and can count knows that six is one more than five and one less than seven. In industrial language, the number six has an *operational definition,* which means that everyone who uses the word *six* agrees on what it means. However, *quality* has no operational definition. That's part of the difficulty in understanding what the quality experts are saying. The other difficulty is that each quality guru thinks his definition is correct and that everyone else's is somewhat flawed. Armand Feigenbaum at General Electric observed that quality doesn't work in bits and pieces; it's either all part of one defined effort or it fails. He calls that single, defined effort *total quality control,* which we will discuss in a later chapter. Kaoru Ishikawa noticed something similar in Japan and also called it total quality control, but it isn't the same as Feigenbaum's effort.

People rarely agree on what the word *quality* means, although they may be talking about the same result. An example was Deming's discussion on quality with Ford Chairman Philip Caldwell in 1981. Deming kept using the phrase "never-ending improvement." Caldwell didn't like the phrase because *never-ending* sounded like too long a time. Deming changed it to "continual improvement," which didn't sound so long. *Continual* and *never-ending* sound different but mean the same thing. The following are other definitions of quality by people who write about and study quality:

- John Stepp said, "I know it when I see it."
- Robert Pirsig wrote *Zen and the Art of Motorcycle Maintenance,* a book that is more about quality than Zen or motorcycles. Pirsig said, "Even though quality cannot be defined, you know what quality is."
- Homer Sarasohn said quality is "a fitness of a process, a product, or a service relative to its intended purpose," which is close to Juran's definition of "fitness for use."
- Crosby defined quality as "conformance to requirements."
- Ishikawa says a quality product is "most economical, most useful, and always satisfactory to the consumer."

Quality must be adopted purposely by every company determined to remain in business. Quality has made the Japanese, Germans, and Koreans economically powerful. However, no single country achieves a perfect grade in quality. Most American companies supply products and services of quality that equals or exceeds their competitors' quality. Quality depends entirely on the way companies operate, on the way their management leads, and on the attitude of the workforce toward their jobs. Even though quality cannot be defined, it can be taught. The most successful companies around the world are acting in ways designed to improve their quality on a continuous basis.

Benefits of a Quality Process

Quality is a pragmatic, dollars-and-cents approach that increases productivity and lowers costs. It is not a subject for philosophical debate, like *beauty* or *justice.* Quality allows larger businesses to profit as they eliminate waste; it allows schools, hospitals, and government programs to run well and not cost an arm and a leg; it helps employees look forward to going to work where they'll be proud of what they do. Quality also means a demanding, difficult, never-ending effort to improve. Some of the unsuccessful attempts at quality failed because the companies started their quality programs without realizing how incredibly demanding they would be. Starting a quality process requires an effort, and maintaining the process after it is running requires a greater effort. It is noteworthy that Japan's Deming Prize winners have demonstrated profit levels twice those of other Japanese companies. Studies in the United States have confirmed the linkage between quality, profits, and market share.

A quality program is not a recipe; it is a way of thinking, and the longer you work at it, the better you get. Instead of solving problems as they arise, a quality program lets you improve the system to prevent problems. Perhaps the most damaging belief that Americans have is that, "If it ain't broke, don't fix it." It sounds sensible, but it is wrong. Why do you want to wait for whatever it might be to break? Why not continually improve the system so that not only does it not break but actually gets better and less expensive? Besides, (1) will you be able to fix it if it does break, (2) how much will it throw your schedule off if it breaks, and (3) will the breakage create a hazard to employees or an irritation to customers? Without a quality process, companies face uphill fights for market share and survival.

THE WAY WE MUST THINK

Every day, members of management and the labor force should question whether they are satisfied with the organization's products and services, and every day they should answer "No." They should never be satisfied with their products and services; they should want to make them better. Look at the evolution of the shaving razor. It started with the straight razor, then the safety razor, then razors with two blades, then the Schick® Quattro® (with four blades), and next the Gillette® Fusion® (with five blades). Someplace, somewhere, people are looking at the Fusion and asking, "How can we make this better?"

As a consumer, I am constantly amazed at the poor quality of some products on the shelves and often shocked at the inadequate service that I receive from customer service representatives. I have become leery of the food raised, packaged, and imported from countries with inadequate environmental and health standards. During Christmas 2007, my wife and I worried that Christmas presents purchased for small children in the family might be tainted with lead paint, and we researched the toys on the Internet to ensure they were safe. For this inconvenience, I blame the company whose name is on the product, not the entire nation where the toy was manufactured. Though the company whose name was on the label did not manufacture the product, a quality-conscious company would have had a valid auditing or inspection system in place to ensure the products were safe for consumers.

American managers are trying to improve things with Six Sigma quality techniques and lean manufacturing. However, I do not believe that lean manufacturing techniques will correct the problems that the United States faces. Plus, American companies will never be lean enough to offset low-cost labor in Africa and the Middle East. No one in this country can afford to work for $5 a day, twelve hours a day, with no benefits. Manufacturing and services need to get back into the reliability game and make things that work a long, long time. If average car buyers keep their cars for ten years, American cars should be so reliable that Americans buy them because they know they will last ten years with minimal problems. Americans will begin "buying American" in greater numbers when American companies pursue and attain the type of quality and reliability indicative of Toyota and Honda.

SUMMARY

Quality is a competitive tool that gives some companies an advantage over other companies that do not have quality systems or that have poor quality systems. Quality is achieved through the sum total of all the systems in a company, not a single system, such as the manufacturing system. To maintain a competitive quality system requires continuous improvement of all systems.

There is no one definition of quality, just as there is no one school or theory of quality. Quality exists in the eyes or heart of the customer. This makes the customer and customer satisfaction extremely important.

REVIEW QUESTIONS

1. Explain what is meant by "Quality is a competitive tool."

2. Explain why quality will be an unending marathon for any business.

3. Write a definition of quality for a riding lawnmower.

4. Explain the advantage of a company having a quality program.

GROUP ACTIVITY

1. Assume you work for a company that has the following sections (systems): a sales force, administration, manufacturing site, raw materials receiving section, and final product loading and shipping section. Pick one section and explain how poor quality in that section would affect the company.

CHAPTER 2

How It All Began

Learning Objectives

After completing this chapter, you should be able to:

- *Explain why companies adopted quality as a business tool.*

- *Explain how General Douglas MacArthur played a part in initiating quality as a business tool.*

- *Explain the roles of Shewhart, Deming, Juran, Sarasohn, and Protzman in the move to quality.*

- *State why the United States resisted the move to quality in the 1950s and 1960s.*

- *Describe the creation of modern industrial production processes.*

INTRODUCTION

You may have heard the word *quality* many times in your life, but have you ever stopped to think about what it means? When you hear the expression *quality product,* what does that mean? When we describe an object as *cool* or *smooth,* meaning that the product is pleasing in some way, we are implying it has a degree of quality. In Chapter 1, you learned that the definition of quality was basically up to the individual or customer.

Interest in and recognition of quality has been a concern since the human race began. Even 4,000 years ago, quality was a concern in building construction and in the safety of a finished building. There was even a severe penalty for not constructing a building safe to occupy. Literally, the law and the penalty for failure were a definition of quality. The following is that code quoted:

> "If a builder builds a house for someone, and does not construct it properly, and the house which he built falls in and kills its owner, then the builder shall be put to death."
>
> *Code of Hammurabi, circa 1780 BCE*

DEFINING QUALITY

In Chapter 1, we briefly looked at definitions of quality. What is the exact meaning of the word *quality?* Quality means different things to different people. It does not necessarily mean that a product is the very best of its kind. For example, individuals buying a new car might have very different requirements of their new cars for them to be considered "quality." See **Table 2.1** for an illustration of this.

If each new car purchaser's requirements were met, then they would consider themselves to have a quality vehicle. They specified what they considered to be a quality vehicle, and the dealership delivered it to them. One good definition of quality is *meeting the customer's requirements.*

Now, let's look at another example of quality in **Table 2.2,** which shows the octane values of regular gasoline at two different gas stations over a period of six weeks. If you add up and

Table 2.1   Quality Requirements for Buying a New Car

Customer	Specifications that Express Quality
Young man	Lots of horsepower, black color, great stereo system, masculine-looking
Young woman	Moderate horsepower, red color, leather interior, roomy, sleek-looking
Elderly woman	Large car, good mileage, comfortable interior, reliable, good resale value
Rancher	Dualie pickup, diesel engine, extended cab, powerful air conditioning

Table 2.2   Octane Values for Gas Stations A and B

Gas Station A	Gas Station B
90	85
90	95
91	70
89	110
90	88
91	92

average their daily octane values, they are the same. Assume they both sell regular gasoline at the same price. At which station would you buy your gasoline?

If you think about it carefully, you will buy gas from Station A. Why? Table 2.2 shows the variations in the octane values for regular gasoline. As you can see, the variation is greater at Station B. Octane value is a measure of the resistance to preignition of a fuel. A certain octane level is required to keep gasoline engines from knocking, pinging, and eventually suffering damage. Octane level also affects gas mileage. This means that, when you fill up at Station B, your gas may not burn the same with each fill-up. With one fill-up, the engine acts zippy; after the next fill-up, the engine pings and runs rough. When you fill up at Station A, your engine always responds the same. Because consumers like consistency and do not like unpleasant surprises, we will probably buy our gasoline at Station A because it has less variation. At Station B, we never know what we're going to get.

The example in octane variation leads us to another definition of quality: *Quality means less variation in the product.* Or, to state it another way: *Quality means consistency.* The phrase *less variation* is a relative one. Less than what? Well, less than a product that has more variation. To create a quality product, we minimize the variation in the specification of the product. Variation in processes is so important that a whole chapter in this book is devoted to it.

Now, if you were the owner of Station B, what would you do to keep attracting customers that are irritated with the inconsistent quality of your gasoline? To stay in business, you may lower your price, and that may work for a short time. But in the long run, your customers may stop coming back if they can't rely on the octane of your gasoline. Lowering your prices even further will not solve the problem. Eventually, your customers will dwindle away and you will go out of business. To state it simply, your business lacked quality management and consistent gasoline quality.

THE BUSINESS NEED FOR QUALITY

Can't you run a business without the consideration and requirements of quality? You can, but you won't be in business for long in today's business environment. Quality has several major business advantages, including the following:

- Quality builds customer loyalty and trust, and retains customers.
- Quality sustains profit margins.
- A better impression helps expand the customer base.
- In the long run, quality will reduce waste and operating costs, and increase profits.

Remember the phrase, "in the long run." Failure to maintain quality consistently over the years has been the downfall of many American companies. As stated in Chapter 1, dozens of large American companies that were in business twenty years ago are not around today.

Quality was a consideration when the first caveman chipped his flint spearhead and mounted it on an ash shaft. He wasn't about to fight a bear with a shoddy spear. Quality was important to the ancient Greeks and Romans and is displayed in their art and science passed down to us. American gunsmiths and silversmiths during the 1700s considered quality in their work. Quality as a concept had always been part of the human race, but it was mostly an individual endeavor. Before the age of mass production, everything was handmade. People had to be craftsmen (those who create or perform with skill or dexterity), because their customers were their own town folk; a reputation and business could be ruined with only a few bad jobs. Mass production—the ability to make a lot of things quickly—began in the early twentieth century, which brought the gradual decline of the craftsman and the beginning of the assembly line worker. Mass production allowed businesses to make a lot of quality things quickly, but also a lot of poor-quality and defective things quickly. The dawn of quality as a critical component of business systems started immediately after World War II.

Nobody stays on top forever. Individuals, corporations, and nations eventually become lazy, conceited, or reluctant to make changes. Rome dominated the world militarily for hundreds of years. Great Britain reigned dominant economically and militarily from approximately 1700 to 1914. After World War II, the United States became the agricultural and industrial supplier to the world. At one point after World War II, the United States controlled one-third of the total world economy and made half of all the manufactured goods sold anywhere in the world. That dominance lasted about fifty years before American economic domination began to yield to Germany and Japan. That is ironic, because prior to about 1970, no one wanted to buy Japanese products, which were deemed shoddy and inferior by Americans. Sometime in the 1970s, Americans began to prefer Japanese and German consumer products over American products. Why? What happened? Well, let's find out.

THE FIRST HERALDS OF QUALITY

If one single person caused Japan's economic turnabout, it was General Douglas MacArthur, Supreme Commander of Allied Forces in Japan after World War II. He was the general who guided the defeat of the Japanese during the war, then unintentionally set in motion a series of steps that morphed Japan into an economic superpower after the war. Japan's economic success was the unintended, serendipitous consequence of MacArthur wanting reliable radios to transmit the orders and propaganda programs of American occupation forces into every town and village in occupied Japan. Because Japanese manufacturers couldn't give MacArthur the quantity or quality of radios he needed, he was forced to send for Americans to teach the Japanese how to manufacture good radios. MacArthur's need for reliable household radios eventually changed the world economic order.

America Teaches Japan

MacArthur sent for Americans Homer M. Sarasohn, a systems and electronics engineer, and Charles W. Protzman, an engineer from Western Electric who joined Sarasohn in 1948, and both started teaching the Japanese how to manage modern manufacturing firms. Their course concentrated on how to manage technology and how to manage a factory in particular. On the first page of their instruction manual, Sarasohn and Protzman quoted an American industrialist named Collis P. Huntington, a tycoon who helped build the transcontinental railroad in the 1860s. At an age when most men retire, Huntington built the Newport News Shipbuilding and Drydock Company in Virginia and wrote the company motto that Sarasohn and Protzman quoted to help the Japanese understand what quality meant:

> *"We shall build good ships here; at a profit if we can, at a loss if we must, but always good ships."*

By this quote, Huntington was saying that they would build ships that customers desired, ships that expressed quality. They would not build inferior products.

Japan After the War

When the war ended, Japan's manufacturing facilities had been nearly destroyed, and it is doubtful the Japanese could have built a single ship, much less ships. American incendiary air raids in March 1945 had turned Tokyo into ashes and cinders. No port city was less than 70 percent destroyed, and no industrial city less than 40 percent destroyed. In August, atomic bombs were dropped on Hiroshima and Nagasaki. Factories were destroyed or closed, public transportation no longer existed, and people were starving. In 1946, when Sarasohn arrived, the Japanese economy no longer existed.

Sarasohn's orders were to build reliable radios to supply the communications needs of the occupation forces and to use the communications industry as an example of how the Japanese economy could be revived. The economic revival of Japan was not a popular idea at the time because many Americans wanted to punish their former war enemy. The logical argument was made that the United States couldn't stay in Japan forever, so the Japanese economy had to be put back on its feet. MacArthur agreed with that argument and ordered the rebuilding of the Japanese industry.

Sarasohn and Protzman, sitting in separate hotel rooms in Osaka for thirty days, wrote their own textbook titled, *The CCS Management Seminar,* consisting of four hundred typed pages. CCS was the acronym for Civil Communication Section. Later, Sarasohn wrote a textbook in Japanese titled, *The Industrial Application of Statistical Quality Control.* Sarasohn wanted to have factories operating before he taught the Japanese any theory. In fact, what he taught them was a complex theory that seemed simple and straightforward to his engineer's mind. He said, "My conception of all of this is that what exists is a system. You're not looking at one factory. You're looking at a *system,* the input of which is your design, the purpose for which you want this item to exist, and everything that it takes to get it to the customer and place that item in his hands to his satisfaction."

SARASOHN'S SYSTEMS APPROACH

The idea that making products or performing services was part of a system had been catching on slowly in the United States since the 1920s and was essentially the second step in the search for quality. The first step had been inspection—the master checking his apprentice's work or the buyer checking the craftsman's product. Inspection works well on an individual basis, but is expensive and wasteful in mass production. In mass production, 500 CD players or flashlights may pass along an assembly line before someone notices they are being made wrong and stops the assembly line to correct the production problem. However, someone still has to be paid to deal with 500 defective manufactured products. In mass manufacturing, an inspector can only separate good from bad after the product is already made, and making a product wrong costs just as much as does making it right (see **Figure 2.1**). About 15 percent to 30 percent of a manufacturing plant's budget goes to building, finding, and fixing defective material if it can be fixed. That is a tremendous amount of wasted resources.

However, if mass production is viewed as a system, then statistics can analyze and control the system. *Webster's Ninth New Collegiate Dictionary* defines a **system** as, "*. . . a collection of different equipment and processes which are interrelated.*" To add clarification to

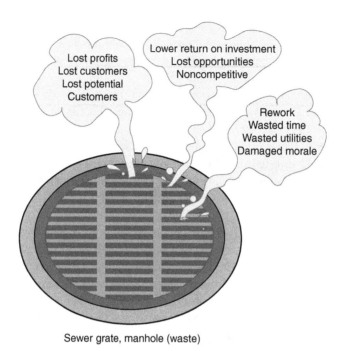

Sewer grate, manhole (waste)

Figure 2.1   The Price of a Lack of Quality

this definition, a system is (1) a group of equipment that works together to accomplish a task, and/or (2) a collection of processes dependent upon one another to complete a task or product. By using a systematic approach, a company, at a very minimum, can eliminate waste, drive down costs, and save money by not having to find and fix mistakes. This systematic concept for quality was being taught to some American engineers during World War II, but the idea did not become widespread.

Sarasohn thought manufacturing should use a systematic approach. He taught this concept to the Japanese. He taught that the production of anything—from a ton of steel, a bank loan, or a restaurant meal—follows a system, and that production cannot be improved unless the whole system is studied to find out what it can do. Those whom Sarasohn wanted to attend were simply ordered to be there; senior executives, company presidents, and chief executive officers attended Sarasohn's CCS management seminar eight hours a day, four days a week, for eight weeks. Some of Japan's leading industrialists became Sarasohn's and Protzman's students, whether they wanted to be or not. They were literally a "captive" audience.

Sarasohn's first question to his students was, "Why is your company in business?" No one had a good answer. That was the starting point for his argument: a company has to have a purpose and a reason for being in business. He said that a company could not be just a moneymaking machine; it had to have a purpose that went beyond mere profit. That purpose might be to make the best and most desired automobile or personal computer. If a company functioned with that purpose, it would always make a profit.

Deming Comes to Japan
Sarasohn wanted quality control to be taught by Walter A. Shewhart, the man who had developed the theory of statistical control of quality while working at Western Electric.

However, Shewhart wasn't available. W. Edwards Deming was invited to lecture in Japan on the new philosophy of quality that had evolved from techniques that he and others had developed for the War Production Board to help improve American war materiel. During World War II, with Deming's help, 35,000 U.S. industrial engineers and technicians were taught to use statistics to determine how to get better results in manufacturing war materiel.

In 1950, Deming taught "Elementary Principles of the Statistical Control of Quality" to 230 Japanese engineers and technicians, using a translator. His lectures were a success because of his ability to persuade senior Japanese managers that he was right about quality. Five years after the end of World War II, Japanese managers learned the management of quality, engineers learned statistical quality control, and the most senior industrialists became aware of the importance of quality. The Japanese began building quality systems into their manufacturing processes.

QUANTITY VERSUS QUALITY

Although Americans were teaching all of this to Japanese engineers and managers, American managers in the United States were busy ignoring quality systems. The common wisdom in the United States was that quality had to be balanced against the cost of attaining it. In other words, quality was another cost factor in production that might not be worth the money spent. This was a common and erroneous concept that prevailed for decades. The idea that higher quality led to lower cost was ignored or unknown in the United States. When Protzman returned to the United States in 1950, he went back to his work at Western Electric (now AT&T) and tried to teach those same principles of quality management to his American colleagues. For his efforts, he was demoted.

Deming said the greatest mistake he made during the war was in teaching quality to American engineers and technicians who design and make products, but not to the management that makes policy. The decision to have a quality system that produces quality products is a policy decision. The design of the process system, the rewards, training, and planning are all part of management policy. The managers who made policy in American business in 1945, decided that quantity was more important than quality.

To be fair, quantity made sense at that time. The manufacturing bases in industrial nations of Europe were severely damaged or destroyed by the war. America had to supply much of the world's material needs and, equally important, had to buy the goods that other nations still produced. In the postwar period, the United States was the engine of world growth and provided a market for the economies of other countries to sell their goods.

At the same time, U.S. industries had to satisfy a ravenous domestic demand for produced goods. The Great Depression was still in effect when World War II broke out. During the Depression, jobs disappeared and people were afraid to spend what money they had. During the war, the economy went to full employment and people suddenly had money to spend, but consumer goods such as tires, gasoline, refrigerators, and automobiles were rationed. Industry had to convert a sizable amount of its productive capacity to war material. General Motors, besides making automobiles, converted some of its assembly lines to making tanks and military trucks and jeeps, as did Ford and Chrysler. A customer had to have a ration coupon to buy a set of tires or a refrigerator.

17

Thus, throughout the Depression and the war, Americans were unable to buy anything except the bare necessities of food and clothes. During the war, Americans had saved $100 billion to help finance the war effort. Now, all that money was waiting to be spent, and it didn't wait long. It financed the demand for houses, cars, and appliances that industrialists could not ignore. Thus, *quantity* was the key to American manufacturing needs at the end of the war. American consumers had piles of money, and they wanted to spend it.

The manufacturing philosophy was to mass-produce it, ship it, and send someone to fix what was poorly made and broke down during warranty. This might be an inconvenience to the consumers, but that was the way it was. Cars couldn't be made fast enough. The concern for quality that had grown during the war all but disappeared as American and world consumers demanded quantity. Thus, *more* is what America had learned to make better than any country in the twentieth century. It was the American economy as much as the Allied fighting forces that won World War II. Both Japan and Germany made some superior weapons, but not enough of them. When President Franklin Roosevelt announced how many bombers were rolling off American assembly lines per month during the war, neither the Allies nor the enemy believed him. Accepted wisdom was that no nation could produce that much, and, in fact, no other nation could. One reason why Japanese Admiral Yamamoto did not want to go to war with America was because he knew the industrial might of America. He knew America could make planes and ships faster than his sailors could destroy them.

In a sense, the United States suffered from the curse of plenty. Although America was very successful at producing quantity, it was on the verge of being a failure because of the lack of attention to quality. Around the early 1970s, American manufacturers realized that they had just begun a marathon race with Europe and Japan and that they were seriously behind in the race. American manufacturers at that time were like the hare in the fable with the tortoise. The hare was racing a turtle (bombed and battered Europe and Japan). Because the tortoise could never beat the hare (or so the hare thought), the hare decided to take a nice long nap. When the hare awoke, the tortoise had reached the finish line. American manufacturers woke up in time to get back in the race, but they had a lot of catching up to do.

MODERN INDUSTRIAL PRODUCTION

Before you can understand quality management and systems, you have to understand mass production. To mass-produce anything, three things are needed:

1. The parts
2. A way to assemble them
3. An efficient way to organize the assembly work

Americans had discovered how to complete all three aspects of mass production. The first machine-made, interchangeable parts were introduced at an industrial exhibition at London's Crystal Palace in 1851. An American gunsmith disassembled ten working rifles, put the parts in a box, mixed them up, and then reassembled them. What we take for granted today was shocking then and so stunning that Europeans referred to machine-made, interchangeable parts as "the American system of manufacturing." Rifles were not mass-produced at that time. Each one was built individually. One man might build the rifle barrel, another the stock, another the trigger mechanism, and so on. And each part, such as the trigger mechanism, wasn't built identical. Thus, on a battlefield, if two rifles were damaged, you couldn't rob parts off of one to repair the other.

Once a manufacturer had succeeded in making parts that were alike, they needed an efficient way to put them together. Henry Ford developed that technique. He knew he could sell every Model T he could make, but he had to devise a way to make a lot of them quickly. He discovered that he could build a lot more cars if he could move the parts past the workers who would remain in one place. His engineers built the first modern assembly line in 1914.

Frederick Winslow Taylor

Once a manufacturer has interchangeable parts and a fast way to put them together, he has to organize people to do the work efficiently. In 1911, the American Frederick Winslow Taylor published what became a classic book, *The Principles of Scientific Management.* The book caused a revolution in productivity. Taylor worked in an iron foundry and began his study by watching an employee shovel sand, and then watching how other workers did their jobs. When he finished his study, Taylor suggested a precise, scientific method of organizing a factory for the most efficiency.

Taylor had very precise ideas about how to introduce his system. He said, "It is only through *enforced* standardization of methods, *enforced* adaption of the best implements and working conditions, and *enforced* cooperation that this faster work can be assured. And the duty of enforcing the adaption of standards and enforcing this cooperation rests with management alone." His unforgivable sin was that he said there was no such thing as "skill" in making and moving things. To the skill-based union members of 1900, this was a direct attack. The introduction of his system was often resented by workers and unions and provoked numerous strikes.

Taylor also dealt with organizing the workforce, reducing each job to its smallest parts and assigning each worker to one repetitive task. With Taylor's method, craftsmanship gave way to boring work done very efficiently. The foundation of modern industrial production now existed. That foundation consisted of (1) interchangeable parts, (2) the assembly line, and (3) scientific management. The economic benefits were quick to show up. In 1908, before the assembly line or Taylor, a Model T Ford cost $850. In 1925, after the assembly line and Taylor, the least expensive Model T cost only $290. Mass production may have nearly eliminated craftsmanship, but it produced modern goods for less money, increased the standard of living in the United States and the industrial world, and completely changed how the world made its money.

Note 1: Taylor's system was extremely productive. Still, the belief in the mystery of craft and skill persisted, as did the assumption that it took long years of apprenticeship to acquire both. Hitler went to war with the United States on the belief of that assumption. Hitler was convinced it took five years or more to train optical craftsmen, whose skills were essential for the optics required in the current warfare. The United States had almost no optical craftsmen in 1941. Using Taylor's methodology, the United States trained semi-skilled workers to turn out more highly advanced optics within a few months than even the Germans were producing.

Note 2: Taylor's work was also the basis for Charlie Chaplin's classic film *Modern Times* and for Aldous Huxley's novel *Brave New World* (which was set in the year 632 AF, meaning *After Ford*). Both condemned the dehumanizing effects of modern technology, but whereas Chaplin and Huxley saw social costs, others saw economic benefits.

AMERICA ENTERS THE UNENDING MARATHON

The Japanese learned how to produce quality from Sarasohn, Protzman, Deming, Juran, and others. Later, the Japanese Kaoru Ishikawa and Genichi Taguchi would add their own quality concepts to Japan's growing body of quality knowledge. Interestingly, no two of these men agreed precisely on how to define quality. In fact, many people think that quality has no one definition, and that the definition of quality can change from generation to generation. The most general definition of quality might be a sense of appreciation that something is better than something else. The difficulty with defining quality is that the definition changes as industry changes. Quality in Leonardo da Vinci's day would not be what it was in Henry Ford's day, or what the definition is in today's digital age. Globally, industrial nations have gone from a craft definition to an industrial definition to what is now called a postmodern definition. Constantly evolving change complicates the definition of quality.

American Management Embraces Quality

During the 1970s and 1980s, quality became an important concept in purchasers' minds. American companies took notice that American citizens seemed to more often find more quality at less cost in foreign products. Lots of cars with names like Honda, Toyota, and Mazda started showing up on our highways. Kids zoomed around on Yamaha and Kawasaki motorcycles. Americans watched Sony TV sets and used Nikon cameras. During those two decades, America was at risk. Simply putting a bumper sticker on your car that said "Buy American!" wasn't going to solve the quality problem, because consumers weren't dumb. They were not going to buy inferior products for the same price as quality products because industry couldn't get its act together. Preferring foreign goods over American-made goods hurt the American economy, created both unemployment and underemployment, and gave America an enormous trade deficit.

At the end of the 1970s, the United States found itself in a quality crisis. Fortunately, American management woke up and suddenly decided that maybe Juran, Deming, and Crosby had something they ought to hear. Clearly, the Japanese had fully embraced the quality techniques of Deming and Juran—techniques that U.S. industry had first developed and then ignored—and Japanese quality was far better than the quality of many American-made products. In 1980, an NBC white paper, "If Japan Can, Why Can't We?" was given credit with starting the U.S. quality revolution. In the early 1980s, American management became serious about producing quality goods and services preferred by people the world over. Though some of Japan's economic success was due to some of its restrictive trade policies, American management realized that the Japanese were not responsible for the American consumer's preference for Japanese products over American products. Management became aware they had problems with customer satisfaction and quality. During the 1980s, a total of 230 American companies—46 percent—disappeared from the Fortune 500 list. Neither their size nor long-established reputations guaranteed continued success in what had become a very competitive world.

America After World War II

The economic control of quality of manufactured product became very widespread during World War II, when the government required statistical quality control (SQC) in contracts with companies producing war materiel. Part of the reason for the requirement of SQC was due to poor welds that caused some Liberty ships in the North Atlantic to break apart and sink. The government did not want to send its fighting men to war with ships that broke apart, airplane engines that suddenly stopped, or explosives that didn't explode.

When the war ended, there was suddenly no regulatory requirement for quality control of materials. American management decided that statistical quality control was a nuisance it could do without. The whole world was America's marketplace ready to buy merchandise. American manufacturing and economic advantages after World War II included the following:

- A huge domestic market
- The world's best technology
- The best educated workforce
- Enormous wealth
- The world's best managers

The economies of all the other industrial nations, except for relatively small Canada and Switzerland, had been damaged or destroyed by the war. Their plants had been bombed and their equipment burned. American managers could sell everything they could make because no one else could make anything or not very much of anything. Thus, American success was guaranteed by world circumstance, not American ability. At his seminars, Deming stated that you had to work especially hard not to succeed in a booming economy with almost no competition.

New Rules

Global markets change, and changes often bring new rules. As the rules for industrial success have changed, so has the way nations calculate theoretical wealth. Natural resources mattered most under the old rules and least under the new rules. What matters today is the production of quality with the only resource that matters—*people*. The United States has always had a treasure chest of natural resources—forests, iron ore, oil, coal—which have given the country an enormous advantage in the world economy. Japan, on the other hand, is a country with few natural resources yet, for a while, Japan economically outperformed the United States.

In the 1990s, Japan and some countries in Western Europe and the rest of the world began to realize that economic advantages don't have to be merely natural advantages (resources). They realized that most of those advantages could be acquired, bought, or manufactured. Whatever technology a country had—supercomputers, machine tools, production processes—could be duplicated or bought by another country, and no country had an advantage based on technology, because that advantage was temporary. Every factor of production other than workforce skills can be duplicated anywhere in the world. Switzerland, Singapore, Japan, Taiwan, and South Korea are small countries almost devoid of natural resources, but each had a well-educated, hardworking, ambitious population. These nations realized that their people were the only true means to create wealth. The key to modern economic success is *human resources*—a well-educated and well-trained workforce that is treated with dignity and respect. Organizations today are always in competition for that most essential resource—qualified, knowledgeable people. Organizations today must attract the right people, and then hold, recognize, reward, motivate, serve, and satisfy them. The scarcest resource in any organization is performing people.

The indisputable fact is that a country needs only to educate and free its people to pursue opportunities. For example, the island of Singapore, a former colony of Great Britain, was once riddled with poverty. Then Singapore embraced education, economic freedom, and foreign investment. By 1985, Singapore's per capita production was approximately 75 percent of Britain's. By the year 2000, the people of Singapore were 30 percent richer than the British who had once ruled them.

SUMMARY

Prior to the advent of mass manufacturing, quality systems were not required; however, once mass production became part of the manufacturing process, quality systems became necessary for a company's survival. A quality system became a company's competitive tool. America was in a unique position after World War II, because it literally had no manufacturing competitors; however, this lack of competition made American businesses complacent. By the 1980s, American businesses awoke to the gradual loss of market share and the customers' desire for quality.

Natural resources are not required for a nation to achieve quality. Indeed, quality can only be achieved with a nation's only true natural resource—its people. A well-educated and well-motivated workforce is the greatest natural resource a country can possess.

REVIEW QUESTIONS

1. Name the person responsible for the economic turnaround of both Japan and the United States, and explain what situation in Japan caused him to inadvertently start the turnaround.

2. List the three things needed to mass-produce anything.

3. Explain what is meant by the phrase, "Inspection works well on an individual basis, but is expensive and wasteful in mass production."

4. The foundation of modern industrial production demanded what social cost and yielded what economic benefit?

5. List two dangers to the American economy if Americans prefer and buy foreign goods over American-made goods.

6. List the five economic advantages America had over its competitors after World War II.

7. Describe four benefits that quality gives to a society or a company.

8. Explain the fallacy of the statement, "If it ain't broke, don't fix it."

MATCHING

1. Sarasohn	A The only resource that matters
2. Shewhart	B "We shall build good ships here; at a profit if we can, at a loss if we must, but always good ships."
3. Deming	C The man who had developed the theory of statistical control of quality
4. General Douglas MacArthur	D Quality doesn't work in bits and pieces; it's either all part of a single, defined effort, or it fails.
5. Collis P. Huntington	E *Principles of Scientific Management*
6. Protzman	F Developed theory of quality for the War Production Board to help improve American war materiel during World War II
7. Henry Ford	G The person responsible for the economic turnaround of Japan
8. Fredrick Winslow Taylor	H An engineer from Western Electric who assisted in rebuilding the manufacturing base of Japan
9. People	I Father of the assembly line
10. Armand Feigenbaum	J A systems and electronics engineer sent to Japan to help rebuild the manufacturing base

GROUP ACTIVITIES

Students should divide into small groups and choose one of the following topics for discussion, then report their conclusions to the class.

1. A company, whether large or small, doesn't need a quality system if it has good, responsive management.

2. Write a response to the "Buy American" bumper stickers that were on so many cars in the 1970s. Would you buy inferior American-made products to save American jobs? Would you buy an American-made product of equal quality if it cost a little more? Explain your responses.

CHAPTER 3

The Quality Gurus

Learning Objectives

After completing this chapter, you should be able to:

- *Explain Deming's philosophy of quality.*

- *Explain Juran's philosophy of quality.*

- *Explain Crosby's philosophy of quality.*

- *Explain the most significant contribution Ishikawa made to the quality improvement process.*

- *Explain the most significant contribution Taguchi made to the quality improvement process.*

- *Explain why a quality system is important to a company.*

INTRODUCTION

In any branch of science or technology, a few people have made significant contributions to their professions. In some cases, they may even have established a new profession or branch of science. For instance, Isaac Newton founded integral calculus. Galileo Galilei and Newton were the physicist gurus of the sixteenth century. Albert Einstein was the key person who changed our views on space and time. Lord William Thomson Kelvin came up with the most important ideas about energy, and said that energy is always the same but can change its forms. Regarding quality, a number of thinkers made significant contributions to the quality profession.

For years, a small group of quality experts had been saying that quality was a cost-effective and necessary business strategy. Five of the most respected quality gurus include the Americans W. Edwards Deming, Joseph M. Juran, and Philip B. Crosby, and the Japanese Kaoru Ishikawa and Genichi Taguchi. All of them recognize that quality has no shortcuts, and that the improvement process is a never-ending cycle requiring the full support and participation of individual workers, and most importantly, the top levels of management. These five gurus disagree about how best to improve quality, but all of their quality systems work; they just have different systems and theories. This chapter introduces you to some of their methodology about achieving greater quality.

W. EDWARDS DEMING (1900–1993)

Dr. W. Edwards Deming was the statistician best known for setting Japanese business on the course that has made them number one in quality throughout the world. Deming was raised by parents who were often poor, hungry, and in debt. He worked odd jobs to help support them, and worked his way through college to earn a degree in electrical engineering. In 1950, he went to Japan to lecture top business leaders on statistical quality control. Deming told the Japanese they could "take over the world" if they followed his advice; he promised that their goods would be world-class in five years. To the Japanese managers, that statement seemed ridiculous because the term "Made in Japan" was such a joke at that time that some factories were set up in a village named "USA," so that products could be stamped "Made in the USA." The rest is history. In 1957, Toyota exported its first car—a real clunker—to the United States; by the 1980s, Japan had become very serious competition to U.S. manufacturing. Deming's sardonic comment was, "Don't blame the Japanese. We did it to ourselves."

Because of what Deming did for Japan, the Japanese have named their highest-quality award the Deming Award. Deming has been called the "founder of the third wave of the Industrial Revolution." The Japanese needed new ways to understand and manage their large industrial processes of the Postindustrial Revolution, and Deming's methodology contributed to that understanding and control. Throughout his life, Deming was also an avid crusader for worker participation in decision making, a concept management did not often accept.

Deming estimated that the United States would need thirty years to accomplish what the Japanese had done to improve quality, because "a big ship (the industrial United States) traveling at full speed requires distance and time to turn." A corporation the size of ExxonMobil or General Motors could not adopt, train, inculcate, and manage a quality system overnight. He warned that managers and people who expected quick results were doomed to disappointment.

Deming's Quality Philosophy

According to Deming, good quality did not necessarily mean high quality. Good quality is a predictable degree of uniformity and dependability at low cost that is suited to the market. The quality of any product or service has many scales and may get a high mark on one scale and a low mark on another. Thus, quality is whatever the customer needs and wants, and, because customers' requirements and tastes are always changing, the solution to defining quality according to customers' viewpoints is to constantly survey the customers.

According to Deming's basic philosophy on quality, productivity improves as variability decreases. Because all things vary, a statistical method of quality control is needed. Statistical control does not imply the absence of defective items; rather, it is a state of random variation in which the ranges of variation are predictable. Deming determined two types of variation: **chance** and **assignable**. Understanding the difference between these was often difficult. He thought that looking for the cause of chance variation was a waste of time and money, yet many companies do this when they attempt to solve quality problems without using statistical methods.

Deming advocated the use of statistics to measure performance in all areas, not just conformance to product specifications. He felt that meeting those specifications was not enough. A company has to keep working to reduce the variation in their processes as well. In other words, a company has to continually improve its product. If a company does not work to reduce variation and a competitor does, the competitor will have a more desirable product.

The Worker and Management

Deming was extremely critical of U.S. management and wanted worker participation in decision making. He claimed that management was responsible for 94 percent of quality problems because management decided policies (who does what, how the production system will work, and who is rewarded). He pointed out that management's task is to help people work smarter, not harder. American management needed to trust and train its workers. Hard work doesn't improve quality or the process; it just makes workers more tired and lowers morale. In the first step of Deming's quality improvement process, management must remove the barriers of a lack of training and proper equipment that rob hourly workers of their right to do good jobs.

Deming disliked motivational programs because he believed that the answer was not simply in everyone doing their best. Motivation did not correct defects in the manufacturing system. Instead, Deming stated that workers cannot do their jobs right when incoming material is off gauge or otherwise defective nor when production equipment is inferior or does not receive proper maintenance. Motivation, which just includes pep talks and posters, is short-lived without constant cheerleading, plus it does not provide the system or tools for workers to do good jobs.

Deming's Red Bead Experiment

In his book *Out of the Crisis,* Deming took issue with the concept of zero defects, deriding it as an example of management sloganeering with little meaning for workers on the shop floor. Part of Deming's argument stems from the idea that rating workers (good worker versus bad worker) is nearly impossible, even if using a bottom-line measurement such as number of defects. As proof, Deming offered his now-famous red bead experiment, in which he filled a large jar with 4,000 beads, 20 percent of them red, the remainder white, all mixed together randomly. Each of six people (workers) puts on a blindfold and draws fifty beads out of the jar. The challenge is to draw out only white beads, which represent "good" products that customers will accept; red beads are considered as defective products that will be rejected. Not surprisingly, the six workers had a large variance in the number of red beads they drew, some having as few as four, some as many as fifteen. It is easy to see that rating workers in this way is foolish. If a process has a certain percent failure rate built into it from the beginning, worker competence is not the issue. If some workers happen to experience fewer defects in a flawed system, that doesn't necessarily mean they are more skilled than other workers. They are just lucky. By the same token, workers who produce more defects aren't necessarily inferior to their colleagues. They are unlucky.

Deming's red bead experiment revealed the fallacy in judging workers in terms of defects, but in a larger sense, it points out management's responsibility for zero defects. Workers may assemble parts or staff call centers, but they're not ultimately responsible for the overall quality direction of the business. Management is responsible for creating processes that work and result in quality products and services.

Vendors

Deming often cited a typical letter from a supplier in response to an inquiry on the supplier's quality. The supplier said, "We are pleased to inform you that quality is our motto. We believe in quality. You will see from the enclosed pamphlet that nothing goes out of this plant until it has been thoroughly inspected. In fact, a large portion of our effort in production is spent on inspection to be sure of our quality." Deming thought the supplier's letter was a true confession of ignorance of what quality is and how to achieve it. According to Deming, inspection, whether of inputs or outputs, is too late, ineffective, and costly. Inspection does not improve or guarantee quality. Inspection merely filters out the good from the bad.

Judging quality requires knowledge of the statistical evidence of quality. Companies dealing with vendors under statistical control can eliminate inspection. Control charts accompanying supplied material can tell buyers what the distribution of quality was, and what it will be tomorrow, far better than any audit. Statistics make quality predictable.

Deming was critical of most procedures for qualifying vendors on quality because, once qualified, vendors are discharged of responsibility, and purchasers accept whatever they receive. The only effective way to qualify vendors is to see if their management abides by Deming's fourteen points, uses statistical process control, and is willing to cooperate on tests and use of instruments and gauges. Deming thought the best recognition for quality vendors is to give those vendors more business. To require statistical evidence of process control when selecting vendors would largely reduce the number of vendors most companies deal with simply because few vendors would qualify (at that time). Deming thought this was the only way to choose vendors, even if it meant relying on a single vendor for critical items.

Deming's Fourteen Points

Deming worked as a private consultant for dozens of clients in the United States and was known to stop working with a client who did not show a total commitment to quality. **Table 3.1** lists Deming's Fourteen Points. When first reading Deming's Fourteen Points, many people find them hard to understand, but keep one thing in mind: Deming focuses on trusting and training the worker. Most people come to work wanting to do a good job and to be proud of their work. This will happen with training and trust.

Table 3.1 Deming's Fourteen Points for Management

1.	Create constancy of purpose toward improvement of product and service.
2.	Adopt the new philosophy. We can no longer live with commonly accepted levels of delays, mistakes, defective materials, and defective workmanship.
3.	Cease dependence on mass inspection. Require, instead, that statistical quality is built in.
4.	End the practice of awarding business on the basis of a price tag.
5.	Find problems. It is management's job to work continuously on the system.
6.	Institute modern methods of training on the job.
7.	Institute modern methods of supervision of production workers. The responsibility of foremen must be changed from quantity to quality.
8.	Drive out fear so that everyone may work effectively for the company.
9.	Break down barriers between departments.
10.	Eliminate numerical goals, posters, and slogans for the workforce, asking for new levels of productivity without providing methods.
11.	Eliminate work standards that prescribe numerical goals.
12.	Remove barriers that stand between the hourly worker and the right to pride of workmanship.
13.	Institute a vigorous program of education and retraining.
14.	Create a structure in top management that will push every day on the preceding thirteen points.

JOSEPH M. JURAN (1904–2008)

Joseph M. Juran was born in Romania and immigrated with his parents to the United States in 1912. Though his parents were poor, he put himself through college through hard work and perseverance. After studying electrical engineering and law, he rose to chief of the inspection control division of Western Electric and professor at New York City University. Like Deming, Juran was credited with part of the quality success story of Japan, where he lectured on how to manage for quality in 1954. He was the author of numerous books on quality. In 1979, he founded the Juran Institute, which continues to this day to conduct quality training seminars.

Juran's Quality Philosophy

Juran believed there were two kinds of quality: fitness for use and conformance to specifications. To illustrate the difference, he said a dangerous product could meet all specifications but would not be fit for use. Juran was the first to deal with the broad management aspects of quality, which distinguishes him from those who advocated specific techniques, such as statistical process control. In the 1940s, he pointed out that the technical aspects of quality control had been well covered, but that firms did not know how to manage for quality. He identified some of the problems as organization, communication, and coordination of functions. In other words: the human element. According to Juran, an understanding of the human situations associated with the job helps to solve the technical problems.

Juran advocated three steps, his ***trilogy of management***, for making progress with quality:

1. Financial planning: Set business financial goals and develop the actions and resources needed to meet those goals.
2. Financial controls: Evaluate actual performance, compare to goals, and take action on the differences.
3. Financial improvement: Do better than the past. Strive for cost reduction and raise productivity.

In Juran's view, workers cause less than 20 percent of quality problems; management causes the other 80 percent.

Just as all managers need some training in finance, all should have training in quality to participate in quality improvement projects. Top management should be included because all major quality problems are usually interdepartmental, meaning members of different departments cause one another unnecessary problems because they are unaware of one another's work processes. Pursuing departmental goals can sometimes undermine a company's overall quality mission.

Juran did not believe motivation campaigns for workforce perfection were effective ways to solve a company's quality problems. A motivational approach fails to set specific goals, establish specific plans to meet those goals, or provide the needed resources. Juran favored quality circles because they improved communications between management and labor. He also recommended using statistical process control but warned that it could lead to a tool-oriented approach to quality.

Juran also recognized the purchasing department's important role in quality improvement. Because a company cannot produce greater precision by itself, it must secure greater precision from its suppliers. The purchasing department's task was more complex than ordinarily assumed. For example, serious consideration must be given to the problems of assessing the quality of contractors competing for big one-of-a-kind projects, as well as how to deal with unexpected changes in specifications. Buyers must recognize the need for better communications with suppliers. This is especially true as more suppliers are foreign firms, which raises the potential barriers of language and cultural differences. Plus, there are also different technological standards throughout the world.

Juran did not favor single sources for important purchases such as raw materials or components. He favored using multiple sources of supply, because a single source can neglect to sharpen its competitive edge in quality, cost, and service. Training for purchasing managers should include techniques for rating vendors. Juran felt that vendors should be trustworthy and part of the manufacturing team.

The Juran Institute teaches a project-by-project, problem-solving team method of quality improvement in which upper management must participate. Juran believed the project approach to be important because there is no such thing as improvement in general when it comes to quality. Any improvement in quality is going to come about project by project, and in no other way. When studying Juran's "Ten Steps to Quality Improvement" in **Table 3.2,** keep in mind that Juran favored looking at the big picture and was the first to deal with the management aspects of quality (trilogy of management).

Table 3.2 Juran's Ten Steps to Quality Improvement

1.	Build awareness of the need and opportunity for improvement.
2.	Set goals for improvement.
3.	Organize to reach the goals (establish a quality council, identify problems, select projects, appoint teams, designate facilitators).
4.	Provide training.
5.	Carry out projects to solve problems.
6.	Report progress.
7.	Give recognition.
8.	Communicate results.
9.	Keep score.
10.	Maintain momentum by making annual improvements part of the regular systems and processes of the company.

PHILIP CROSBY (1926–2001)

Philip Crosby is the quality expert best known for developing the concept of zero defects in the early 1960s when he was in charge of quality for the Pershing missile project at the Martin Corporation. In 1965, he went to ITT as director of quality and left in 1979 to form his own company, Philip Crosby Associates. He got into quality consulting and writing because he was tired of hearing how the United States was going downhill.

Crosby's Quality Philosophy

Crosby defined quality as conformance to requirements and said it could only be measured by the **cost of nonconformance**. He didn't believe in talking about poor quality or high quality. Instead, he talked about **conformance** and **nonconformance**. Using this approach meant that the only standard of performance was **zero defects**. Even today, the quest for zero defects is still considered quality's holy grail. Crosby firmly believed industry could achieve zero defects. In all industries, defects cost money, waste time, and frustrate managers. In some industries, such as pharmaceuticals or medical devices, production errors can cost lives.

If he had to use a single word to define what quality management was all about, Crosby chose the word **prevention**. In the 1970s and '80s, the conventional view was that quality was achieved through inspection, testing, and checking. Crosby believed that prevention was the only system that could produce quality, and that prevention means perfection. He did not believe in statistically acceptable levels of quality, because people go to elaborate measures to develop statistical levels of compliance, and eventually come to plan for error because they believe that it is inevitable. Crosby believed there was absolutely no reason for having errors or defects in any product.

Crosby pointed out that quality improvement was a process, not a program, because nothing permanent comes from a program. Quality was management's responsibility, and management had to be as concerned about quality as production, safety, and profit. It bothered him that most companies continued to compound their quality problems by treating their

31

employees poorly. The thoughtless, irritating, and unconcerned way employees were often treated was very de-motivating. Crosby maintained that a committed management could obtain a 40 percent reduction in error rates very quickly from a committed workforce.

Crosby also believed that the purchaser caused at least half of the quality problems associated with purchased items by not clearly stating what the requirements were for the items. Because defects are defined as deviations from the agreed-upon requirements, a lot of effort and thought should go into requirements. Crosby used Japan as an example, where Japanese firms treated suppliers as extensions of their own businesses. Half of the rejections that occurred in American companies were the fault of the purchaser. He also believed that visiting a potential supplier to conduct a quality audit was next to useless. Knowing whether a vendor's quality system will provide the proper control or not is impossible unless the vendor site is a complete and obvious disaster area.

The Cost of Quality

In most cases, the biggest factor in a company's quality program is cost. The chief business of business is business. In other words, successful quality programs must represent money to the companies that undertake them—money that they either earn or save. The pursuit of zero defects is justified only if it satisfies this requirement, the most important quality metric of all.

Philip Crosby asserted that "quality is free" in his book of the same name, which brought the concept of zero defects into the mainstream. Most managers agree that poor quality costs a great deal. Inefficient manufacturing processes can lead to scrapped materials and necessitate time-consuming and expensive rework. But preventing this through inspection and corrective action is also costly. This was recognized at least as far back as the 1940s, when early quality guru Armand Feigenbaum began to express inefficient manufacturing processes in the form of dollars lost. Crosby furthered this idea by pointing out that preventing defects in the planning stages of a process is the cheapest and surest way to ensure quality. Flawed processes (systems), designed by leaders who haven't fully worked out the ramifications of those flaws, are the ones most likely to lead to defective products and high inspection or rework costs. Managers set the tone for change within their working environments. Workers on the shop floor follow the rules and behaviors set forth by their leadership. No amount of worker conscientiousness, dedication, or foresight will bring about a zero defects state in the absence of management commitment to creating processes that work correctly.

Zero Defects in the Real World

Human beings aren't perfect, but our processes and systems can approach perfection even if we can't achieve that state ourselves. The existence of human error is the reason why we install systems with alarms or warning devices for our most important tasks. Crosby believed it was incumbent upon managers to create processes that lead to defect-free work. The best way to reduce manufacturing costs is to perform tasks the right way the first time. But zero defects isn't about perfection; it is about the understanding on the part of every member of the organization that processes must constantly be improved, and that defective systems must be reworked and reorganized from the top down. The zero defects philosophy is an attitude and a performance standard. With a zero defects mind-set, each defect is rigorously traced to its **root cause**, and each cause is prevented (eliminated). Inspection cannot lead to zero defects. In the zero defects philosophy, inspection isn't just a goal; it's a requirement that leads an organization toward the zero defects state. Management asks, "Are we completely eliminating this problem forever?" and the organization searches for ways to reply, "Yes."

For those who think that the quest for zero defects is impossible, too difficult, or too expensive a goal, Crosby had a ready answer: Defects are not inevitable and they are not acceptable. As he wrote in *Quality Is Free,* "People are conditioned to believe that error is inevitable. We not only accept error, we anticipate it. Whether we are designing circuits, programming a computer, planning a project, typing letters or assembling components, it does not bother us to make a few errors and management plans for these errors to occur."

However, we do not maintain a standard of accepting errors when it comes to our personal life. If we did, we would resign ourselves to being shortchanged now and then as we cash our paychecks. We would expect to be given the wrong medication from the pharmacy. We would expect to be arrested for a crime we didn't commit. As individuals, we do not tolerate these things. Thus, we have a double standard—one for ourselves, and one for the company job. Defects are not normally accepted events in our personal lives; there is no reason that they should be accepted normally in service or manufacturing organizations.

Crosby's Absolutes of Quality

Crosby created the following four principles that he called the "Absolutes of Quality":

- *Quality is defined as conformance to requirements.* Everyone has to have and understand the definition of quality. If the requirement is to paint a car jet-black with no paint runs, and you do that, then you have a quality paint job.
- *The system that will make quality happen is prevention.* Prevention is eliminating the potential for defects, nonconformances, or error. Prevention is identifying opportunities for error and taking action to keep the errors from occurring.
- *The performance standard is zero defects.* A performance standard specifies how often you want something (the process) done right. If your performance standard is for 97 percent done correctly, then you allow the potential for 3 percent defects into your process and a worker attitude that accepts some defects.
- *The cost of quality is measured by the price of nonconformance (PONC),* which is measured in dollars. This measurement of dollars is that wasted on rework, reprocessing, change orders, downtime, revisions, and so on.

Crosby also had fourteen steps for achieving quality and these are listed in **Table 3.3.**

Crosby's quality system differs from Juran's and Deming's in that it focused a lot on people and their expectations. He believed strongly in motivation and personal behavior.

THE JAPANESE GURUS

The Japanese have made significant contributions to quality concepts and methodology. They have worked on quality methods with the dedicated passion of religious zealots. The Union of Japanese Scientists and Engineers (JUSE) and the Japan Standards Association run hundreds of courses yearly for engineering students. The core of all the courses is total

Table 3.3 Crosby's Fourteen Steps to Quality Improvement

1.	Make it clear that management is committed to quality.
2.	Form quality improvement teams with representatives from each department.
3.	Determine where current and potential quality problems lie.
4.	Evaluate the cost of quality and explain its use as a management tool.
5.	Raise the quality awareness and personal concern of all employees.
6.	Take action to correct problems identified through previous steps.
7.	Establish a committee for the zero defects program.
8.	Train supervisors to actively carry out their part of the quality improvement program.
9.	Hold a "Zero Defects Day" to let all the employees realize that there has been a change.
10.	Encourage individuals to establish improvement goals for themselves and their groups.
11.	Encourage employees to communicate to management the obstacles they face in attaining their improvement goals.
12.	Recognize and appreciate those who participate.
13.	Establish quality councils to communicate on a regular basis.
14.	Do it all over again to emphasize the quality improvement program never ends.

quality control (TQC), based on statistical methods. From 1960 to 1985, JUSE published 660 books on quality control, and it publishes three magazines a month on quality. It would not be an exaggeration to say that Japan has more information on quality control than any other country.

Two of the best-known Japanese quality gurus are Kaoru Ishikawa and Genichi Taguchi. Their methodology and concepts have been adopted and used extensively worldwide.

KAORU ISHIKAWA (1915–1989)

As a professor of engineering at Tokyo University just after the war, Kaoru Ishikawa is the best known of the Japanese quality experts, unlike Deming and Juran, who were poor. Ishikawa graduated in 1938 from Tokyo University, which is the Japanese equivalent of Harvard. Ishikawa got into the quality movement unintentionally, as did several of the American engineers sent to Japan to help rebuild their manufacturing bases after the war. One of Ishikawa's best-known works is his title, *What Is Total Quality Control?*

By 1949, Ishikawa was urging statistical methods of quality control at the JUSE. Many people think his life and the history of quality control in Japan cannot be separated. When American executives began going to Japan in the early 1970s to learn how its industries produced better quality at lower cost, many of them came back with Ishikawa's innovation, the **quality control circle**. A quality control circle consisted of small groups of Japanese workers who met voluntarily to discuss ways to improve their own work and to make suggestions for improving their manufacturing system. The circles met in the plant, and they were unlike anything in the United States. American executives somehow believed those quality control circles were the answer to their quality problems and they imported the idea.

Quality Control Circles

Ishikawa developed quality control (QC) circles in 1962, and persuaded Japanese management to not only support but to listen to the QC circle suggestions as part of the total quality effort. That was the key difference between American and Japanese quality control circles: Japanese management listened to the workers. American workers made suggestions every bit as good as their Japanese counterparts, but American managers usually didn't listen to them. In the United States, quality circles—the word *control* was dropped—were popular in the 1970s and early '80s, but often died due to a lack of the total quality program that existed in Japan. Because total quality is largely an attitude and mental process, American visitors to Japanese plants could not see it and therefore did not import it.

Crosby remembers Americans going to Japan and coming back enthusiastic about quality circles; he said that quality circles were a great idea but were ineffective alone because, though workers tried to change and wanted change, they soon found out that management had not changed its myopic ways. Crosby said that American managers jumped on quality circles because the concept put the burden of quality on the workers, not on the managers. Crosby knew that if management would not commit to quality, then dedicated workers would be unable to help the quality improvement process.

When Juran taught his quality improvement courses to Japanese executives, the courses presented definitive answers to detailed questions posed by Japanese companies. These lectures led the executives to feel that a pathway back to prosperity was suddenly available to them. They were enthusiastic and ready to try new approaches to developing effective means of Japanese TQC. Ishikawa felt that Juran's visit created an atmosphere in which QC was to become a tool of management, creating an opening for the establishment of TQC as it is known today.

Ishikawa and Crosby

If Ishikawa and his quality improvement method are compared with Crosby and his method, we will find that they agree on the following:

- Top management must provide continuous support and leadership in the quality improvement process and act as role models to employees.
- Management must set quality standards and measure performance.
- Management is primarily responsible for poor quality.
- Management must commit to continuous learning for the workforce.
- Quality improvement leads to cost reduction.

The principle difference between Crosby and Ishikawa is that Crosby's methodology stresses the importance of showing management the cost of not producing quality products and services. Ishikawa's methodology stresses that all the workers (not just the managers) must work on quality and know why quality is important. In Ishikawa's method, the cost of nonconformance is revealed to all employees.

The Internal Customer

In the core of Ishikawa's total quality control methodology, he places primary importance on the role of the customer, but he changed the definition of customer. While working with a Japanese steel mill in 1950, Ishikawa came up with the idea that the customer is not necessarily the person who buys the product or pays for the service. The customer is the next worker in line to receive your work, and who depends on you to have done your part correctly. The customer is the next person on the assembly line, the clerk who makes out the invoice, the salesman. That occurred to Ishikawa at the steel mill when he heard workers grumbling about things going wrong in different areas that would affect them. Ishikawa asked the workmen in that section of the steel mill why they didn't go to the next section and ask the workers there what was wrong. The workmen said they couldn't do that because those workers would think they were spying on them and would run them off. This made Ishikawa aware that company-wide quality control could never occur with this kind of worker viewpoint. It gave him the idea of the new customer. Departmentalism had to be broken down. That is the spirit of TQC. Quality experts now widely accept the identification of coworkers and colleagues as customers. For this, Ishikawa won the Deming Prize in 1952.

GENICHI TAGUCHI (1924–)

Like Crosby and Feigenbaum, Taguchi tried to place a value on the cost of lack of quality, but his methodology places it in each step of the process, from a product's design to its sale and use. He maintains that quality is primarily a function of design, and that the strength of a product's quality is primarily the responsibility of the product designers. Good factories are faithful to the intention of the design. Mediocre designs will always result in mediocre products. One of the things that the Japanese see very clearly is the loss to society that comes from non-quality. That is implicit in the Taguchi loss function, which potentially has its roots in some of Dr. Deming's earlier teachings.

However, Deming and Taguchi appear to differ on the cost of a lack of quality. Deming and Taguchi agree on the loss function itself; they disagree on whether it shows what the loss actually is. Deming says the actual loss due to a lack of quality can't be calculated; Taguchi says the cost of non-quality can and should be calculated. Although the mathematical theory may be complex, the loss function is fairly easy to understand, at least in concept.

Taguchi recognizes an incremental economic loss for any deviation. This view is quite different from the traditional view that there is no loss so long as the parts are within the engineering tolerances (specifications). Traditionally in industry, parts are made to certain specifications. For example, for a widget that should ideally be 500 millimeters wide, the design engineer designs an engineering tolerance (specification) of ± 5 mm (495 to 505) and declares that good enough.

The Taguchi Loss Formula

The practical application of the Taguchi loss formula can save an enormous amount of money. Decades ago, the Ford Motor Company had trouble with a particular transmission. Those built in the United States were breaking down and causing a huge increase in warranty costs. The same transmission built to the same specifications in Japan had no problems. Engineers took apart twelve transmissions from the United States and twelve from Japan. Every piece of every American transmission was tested and measured, and every single piece was within specification. Nothing was wrong with the American-built transmissions. They were what they should be, except for their unfortunate tendency to break down. But Taguchi maintained that, when enough parts were near the outer limits of their specifications and assembled in one transmission, all those trivial deviations had a magnifying effect on one another. Every part was within specification, but so many were at the outer limits of specification, that, as a group, they caused transmissions to fail.

When the Japanese transmissions were tested, the engineers reported that the measuring device had broken. It had not. There was so little variation from part to part that every part was essentially the same. Referring back to our widget, 495 millimeters is not the same as 500, but 499.9 millimeters could be so close that it wouldn't matter. From the Taguchi loss function, the Japanese realized that their transmissions were better the closer they got to the ideal, and they ignored the outer limits of the specifications and worked to continually reduce variation toward the perfect or near-perfect. That is the point of statistically based quality programs: the less variation, the greater the quality. Deming maintained that achieving that ideal figure was not necessary, that you probably wouldn't know if you did, and that the goal is not the achievement of perfection but the reduction of variation. Taguchi realized any deviation from the ideal led to incremental losses, some hard to measure, but still measurable.

SUMMARY

There are several philosophies and schools of quality, just as there have been and still are founders of specific quality concepts. They offer different paths leading to the same destination—a service or product of excellence, dependability, and desirability. Some schools stress data collection and statistics, others training and worker empowerment, and still others a systems approach versus incremental improvement. The important thing is that corporations adopt a quality system for their competitive advantage.

Everything we need to know about quality has not been discovered. Different schools of philosophy will continue to emerge, different tools for managing quality will be developed, and quality as a competitive business tool will continue to evolve.

REVIEW QUESTIONS

1. List the three most respected American quality gurus.

2. Explain what Deming's quality system focused on most.

3. Explain what Juran's quality system focused on most.

4. Explain what Crosby's quality system focused on most.

5. Describe the most significant contribution Ishikawa made to the quality improvement process.

6. Explain the most significant contribution Taguchi made to the quality improvement process.

7. Who did Juran, Crosby, and Deming claim was responsible for about 89 percent of all quality problems?

8. (True or false) Zero defects means the worker comes to work each day and must not make a single mistake.

9. How did Crosby define *zero defects?*

10. What was Deming saying with his red bead experiment?

11. What three things did Ishikawa contribute to the study of quality?

12. Taguchi is recognized for what contribution to the study of quality?

13. From Taguchi's viewpoint, explain what is wrong with building things merely within tolerance.

GROUP ACTIVITIES

Students should divide into small groups and choose one of the following topics for discussion, then report their conclusions to the class.

1. List reasons why you agree or disagree with Deming's tenth point, "Eliminate numerical goals, posters, and slogans for the workforce, asking for new levels of productivity without providing methods."

2. List reasons to support or deny Crosby's first step, "Make it clear that management is committed to quality."

3. Explain why you either agree or disagree with Crosby's concept of zero defects.

CLASS DISCUSSION MATERIAL

Discussion Item #1—The following are seven scenarios. Determine if they were "quality" experiences based on Philip Crosby's definition of quality, which is "conformance to requirements." Remember, the provider of the product or service needs to deliver it in a manner that meets the requirements of its customers. Your instructor has the answers.

SCENARIO

1. You buy your son a water gun at a store for one dollar. He plays with it for two hours before the trigger breaks. It no longer squirts water, but he continues to play with it.

2. You buy a $40,000 SUV. Within twelve months, the air conditioner needs to be repaired and the power mirrors need to be replaced. You take it into the dealership and they fix everything for free and loan you a free rental car.

3. You buy a $24,000 Toyota Camry. After signing all the paperwork, you notice a small scratch on the hood. The dealer promises to make it as good as new. After the repair, you can't tell that the scratch was ever there.

4. You check into a four-star hotel in New York. Your $299-a-night room is on the 27th floor. You have to wait about eight minutes for the elevators each time you want to go up or down.

5. You check into a $49-a-night room at a Motel 8. There is no elevator, so you have to carry your luggage to the second-story room. The room is clean and sparsely furnished.

6. You go through the drive-through at McDonald's at noon and order a salad with grilled chicken. You are told that you have to park and wait ten minutes while your chicken is cooked. Your meal is delicious.

7. You buy a new computer from a warehouse club retailer. The computer is plagued with problems, and you eventually return it for a full refund after eight months.

Discussion Item #2 Read the following newspaper article, and discuss the system failures using Deming's Fourteen Points for quality management.

TWO INFANTS DIE AFTER GETTING ADULT DOSES

(This is a paraphrased report of a newspaper article. Names of people and places have been omitted or changed.)

INDIANAPOLIS (AP) Two premature infants died after receiving adult doses of a blood thinner, a hospital said Sunday, blaming the incident on human error. "Four other infants in the neonatal intensive care unit of Methodist Hospital also received adult doses of heparin, and one might need surgery," said Sam Doe, chief executive of Methodist and Iowa University Hospitals.

The other three infants were in serious condition. Two babies born at 25 and 26 weeks gestation died Saturday night. Both were born in the last week, officials said. A full-term pregnancy lasts 38 to 42 weeks.

"These are very, very small babies," Doe said. "We are confident that no other infants except for the six were affected."

Heparin is routinely used in premature infants to prevent blood clots that could clog intravenous drug tubes. An overdose could cause severe internal bleeding. Hospital officials had met with family members, saying that their hearts went out to the families. But apologies did not satisfy Jane Doe, mother of one of the infants who died. "They may apologize but it didn't help," she told a TV reporter in Indianapolis. "It didn't help, because I feel like whoever the nurse was on call, they should have known what they were doing and how much my baby should have."

The hospital was investigating how the error occurred and reviewing its drug-handling procedures. Some corrective steps had already been taken. "This was human error, that's all," Sam Doe said. Doe said the drug arrives at the hospital in premeasured vials and is placed in a computerized drug cabinet by pharmacy technicians. When nurses need to administer the drug, they retrieve it from a specific drawer, which then relocks. Somehow, the adult doses were placed in the drug cabinet in the neonatal intensive care unit. There was no evidence that infant doses were given to any adult patients. The adult and infant doses are similarly packaged. The hospital will ask the manufacturer to make the packaging more distinct.

39

CHAPTER 4

The International Standards Organization

Learning Objectives

After completing this chapter, you will be able to:

- *Describe the history of the International Standards Organization (ISO).*

- *Explain the purpose of ISO registration.*

- *Explain the benefits of ISO registration.*

- *List some of the documents that must be in place for ISO registration.*

- *Explain the differences between TQM and ISO.*

- *Explain how ISO 2000 is different from ISO 9000.*

INTRODUCTION

The 1980s was the decade of the quality gurus. Their ideas are still valid, and their colleges and institutes still exist; however, in the 1990s, many businesses and corporations embraced a quality system referred to by the acronym **ISO**, which stands for the International Organization for Standardization. Many of the companies had already subscribed to a quality system, either Deming's, Juran's, Crosby's, or others, but then decided that the quality system of the future looked like it was going to be ISO.

For many of the companies, ISO was a necessary business decision. The ISO 9000 system of today is in the process of revision and morphing into ISO 9001-2000. This will be discussed later in this chapter. A general relationship of the ISO 9000 standard to other quality standards or awards is shown in **Figure 4.1.**

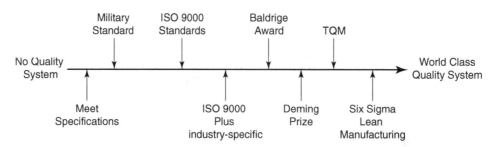

Figure 4.1 ISO 9000 Standard Relationship to Other Quality Standards/Awards

PRODUCT AND SERVICE STANDARDS

A standard is something widely used, well established, and acceptable. Mankind cannot exist as a communal species without standards. We would soon notice if standards we are unaware of were to disappear. Standards make an enormous contribution to most aspects of our lives—although that contribution is very often invisible. The importance of standards is brought home only when standards are absent. For example, as purchasers or users of products, we soon notice when they turn out to be of poor quality, do not fit, are incompatible with equipment we already have, or are unreliable or dangerous. When products meet our expectations, we tend to take this for granted. We are usually unaware of the role standards play in raising levels of quality, safety, reliability, efficiency, and interchangeability—as well as in providing such benefits at an economical cost. Without standards, the world would be much more unsafe, things would be much more unreliable, and efficiency would be greatly reduced.

The Geneva-based International Organization for Standardization (ISO) first published a series of standards in 1987. Known as the ISO standards, they provide a basis by which companies achieving ISO registration would assure buyers that they have a quality system in place that produces goods to customer specifications. Originally, in the early 1990s, TQM and ISO had the following differences:

- ISO standards were created only to improve a firm's production processes.
- A TQM system is the big picture and is concerned with customer satisfaction and all activities conducted by a firm.

Since then, ISO has evolved and made gradual changes, which will be discussed later in this chapter.

The ISO standard's emphasis is on the management of *process quality,* assuring that the production process creates material according to specification and is managed so as to continue manufacturing product according to specification all the time. ISO provides an excellent beginning platform for a firm wanting to go further into quality.

Almost everyone working inside the fence of a chemical plant or refinery will have a part to play in an ISO quality system. They will receive training in their company's ISO standard and will be expected to help maintain the plant ISO quality system. They will be required to be important members of the ISO quality team.

HOW ISO BEGAN

In 1946, delegates from 25 countries met in London and decided to create a new international organization with the objective "to facilitate the international coordination and unification of industrial standards." The new organization, ISO, officially began operations on February 23, 1947. However, progress was very slow, and Europe had not yet discovered how to make ISO a competitive tool that would benefit it.

In the 1980s, the twelve countries in Western Europe that made up the European Union (EU)—Germany, France, Spain, Italy, and so on—agreed that they could become more competitive and efficient if they created something like the United States of Europe. By pooling their populations and landmass, and reducing and eliminating tariffs and trade barriers, these nations could become a tremendous economic powerhouse. ISO was one tool these nations would use in competing against the economic giants of the United States and Japan.

To meet its objective of creating a single unified market by the end of 1992, the twelve-nation European Union implemented over 300 directives that established new regulations covering a broad range of business activities and thousands of products. Designed to ease the free flow of goods and capital within the single European market, these regulations substantially altered the way in which U.S. companies who exported to Europe conducted their business. ISO 9000, the common name for a set of five standards sponsored by the International Organization for Standardization, establishes requirements with which a company's quality management system must comply to export competitively into the European Union.

The standard has proliferated across a broad spectrum of organizations. Twenty years ago, few would have thought that financial institutions, schools, blood banks, prison systems, Buddhist temples, cruise lines, or retail chains would implement ISO 9001 compliant quality management systems. No one foresaw its success or the acceptance it has gained around the world.

The Purpose of ISO

ISO, with over 90 member countries, was created to develop and promote standards of all sorts worldwide. ISO work is done through over 180 technical committees. ISO/TC176, the Quality Management and Quality Assurance Technical Committee, is responsible for quality standards. The U.S. representative to ISO is the American National Standards Institute (ANSI). Because "International Organization for Standardization" would have different abbreviations in different languages (IOS in English, OIN in French for *Organisation internationale de normalisation*), it was decided at the outset to use a word derived from the Greek *isos,* meaning "equal." Therefore, whatever the country, whatever the language, the short form of the organization's name is always *ISO*.

ISO 9000 is a series of five generic, baseline quality standards (**Figure 4.2**) intended for broad application to a wide range of nonspecific industries and products. These standards define the basics of how to establish, document, and maintain an effective quality system. The ISO 9000 series consists of (1) models that define specific minimum requirements for external suppliers, and (2) guidelines for the development of internal quality programs.

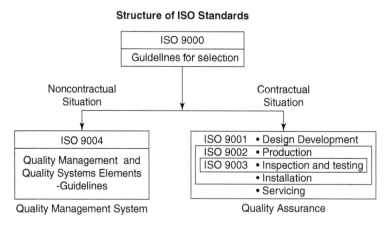

Figure 4.2   Structure of ISO Standards

ISO standards are voluntary. As a nongovernmental organization, ISO has no legal authority to enforce their implementation. A certain percentage of ISO standards—mainly those concerned with health, safety, or the environment—has been adopted in some countries as part of their regulatory framework. These nations have decided that some materials and equipment must meet ISO standards to be manufactured in the country or imported into the country. Such adoptions are sovereign decisions by the regulatory authorities or governments of the countries concerned. ISO itself does not regulate or legislate. However, although ISO standards are voluntary, they may become a market requirement.

ISO develops only those standards for which there is a market requirement. The work is carried out by experts from the industrial, technical, and business sectors that have requested the standards, and that subsequently put them to use. These experts may be joined by others with relevant knowledge, such as representatives of government agencies, consumer organizations, academia, and testing laboratories.

Although ISO standards are voluntary, they are developed in response to market demand and are based on consensus among the interested parties, ensuring widespread applicability of the standards. Like technology, consensus evolves, and ISO takes account both of evolving technology and of evolving interests by requiring a review of its standards at least every five years to decide whether they should be maintained, updated, or withdrawn. In this way, ISO standards retain their position as the state of the art, as agreed by an international cross section of experts in the field.

ISO as a Quality System Standard

ISO 9001, one of the quality standards, is a not a product standard, but a quality system standard that applies not to products or services, but to the process that creates them. Its standards can consist of as many as twenty elements and are designed to apply to practically any product or service made by any process anywhere in the world. The elements that make up the standard consist of requirements for document control, process control, inspection and testing, purchasing, corrective action, and so on. Each element is written in very general terms that specifically describe *how* to accomplish the specific objectives of the standard. The standards are written to a generic state, avoiding as much as possible the mandating of specific methods, practices, and techniques. Instead, they emphasize principles, goals, and objectives that focus on one primary objective critical to every manufacturer: *meeting customer expectations and requirements.* Although ISO 9000 is not perfect, it remains the world's best-known and most-used standard.

STANDARDIZATION AND BENEFITS

When the large majority of products or services in a particular business or industry sector conform to International Standards, a state of industry-wide standardization is said to exist. This is achieved through consensus agreements between national delegations representing all the economic stakeholders concerned—suppliers, users, government regulators, and other interest groups such as consumers. They agree on specifications and criteria to be applied consistently in the classification of materials, in the manufacture and supply of products, in testing and analysis, in terminology, and in the provision of services. In this way, International Standards provide a reference framework or a common technological language between suppliers and their customers. This facilitates trade and the transfer of technology.

Benefits of International Standards

For businesses, the widespread adoption of International Standards means that suppliers can base the development of their products and services on specifications that have wide acceptance in their sectors. This, in turn, means that businesses using International Standards are increasingly free to compete in many more markets around the world.

For customers, the worldwide compatibility of technology that is achieved when products and services are based on International Standards brings them an increasingly wide choice of offers. Customers also benefit from the effects of competition among suppliers.

For developing countries, International Standards that represent an international consensus on the state of the art constitute an important source of technological know-how. By defining the characteristics that products and services will be expected to meet in export markets, International Standards give developing countries a basis for making the right decisions when investing their scarce resources and thus avoiding squandering resources.

For consumers, conformity of products and services to International Standards provides assurance about their quality, safety, and reliability.

For the planet we inhabit, International Standards on air, water, and soil quality, and on emissions of gases and radiation, can contribute to efforts to preserve the environment.

AMERICA EMBRACES ISO

In the early 1990s, American firms reacted to ISO 9000 in various ways. Though some were skeptical and wondered just how the standards would benefit them, the answer was obvious to quite a few firms. Without ISO 9000, they were threatened with a loss of business due to the following:

- The European Union (EU) might make ISO 9000 registration mandatory for American firms in their marketplace.
- The American firms' biggest customers might require ISO 9000 registration of all suppliers.
- Their major competitors might adopt ISO 9000.

A huge number of American firms were drawn into the ISO registration orbit for just one of the many practical benefits: protecting their existing markets.

The ISO 9000 quality system standard has tactical and strategic applications because ISO objectives affect competitiveness as well as quality. For instance, a process industry with an effective quality system will (1) strengthen its own competitiveness, and (2) achieve product quality in a cost-effective way. ISO 9000 is neither new nor radical. Like TQM, the ISO 9000 is good, hardheaded business sense. Companies that have achieved ISO certification repeatedly describe the following six benefits:

1. A more efficient and effective operation
2. Increased customer satisfaction and retention
3. A reduction in audits
4. Enhanced marketing
5. Improved employee motivation, awareness, and morale
6. Promotion of international trade

The European Union Council of Ministers has mandated ISO 9000 certification for makers of certain types of products, such as commercial scales, construction products, gas appliances, industrial safety equipment, medical devices, and telecommunications terminal equipment. More products and services may be added to this list, especially those that are potentially hazardous, that involve personal safety, or that are affected by product liability. Mandated ISO 9000 registrations are not common. The pressure for ISO registration is principally driven by nongovernmental forces, such as some major firms in Europe and elsewhere that mandate ISO 9000 registration for their suppliers. Suppliers are warned that they must be registered by a certain date and are normally given plenty of time to attain registration. For firms put in this situation, certification enables them to keep their existing customers.

Most of the pressure for ISO registration is not in the nature of competitor threats or mandated registration. Instead, it is growing international competition and the needs of firms everywhere to have access to markets. This is a critical benefit of ISO 9000 certification because it separates good from poor suppliers of materials.

A final benefit of ISO 9000 registration is the potential for reduced customer audits. Many facilities in certain industry segments undergo dozens of quality audits each year. These audits take managers, operators, chemists, clerks, and everyone involved in the audit process away from their daily duties to be available for the auditors. Those people are no longer doing their daily jobs. This can affect productivity and, if customers conduct too many audits, morale may be affected. Some companies have as many as three to five customer audits a month. Now that ISO 9000 registration is understood and accepted in the United States, many customers accept current ISO 9000 registration in lieu of site audits. They know the registrars' **surveillance audits** force their suppliers to maintain effective quality systems.

APPLYING FOR AN ISO STANDARD

Though there are five ISO standards, ISO 9000 is an overview and ISO 9004 is used for internal quality, not for outside certification (registration). Thus, a firm must first decide which of three levels of certification to pursue: ISO 9001, 9002, or 9003, which are quality system models. The facility becomes certified to the model that most closely fits the scope of its operations, services, manufacturing, or retailing. The facility's quality system includes only those elements of the standard that are relevant to the effective operation of the facility. The quality system must be documented through one or more levels of documentation, including, for most facilities, a *quality manual* as the top level of documentation. Companies choose the level of certification based on the following:

- If a firm is engaged in design efforts and/or after-sale servicing, the firm typically seeks ISO 9001 registration.
- ISO 9002 registration is for firms involved in production and installation, but that produce to other firm's specifications.
- Distributors and retailers typically seek ISO 9003.

To a great extent, the range and intensity of the ISO 9000 benefits are determined by the way the standard is applied. The standard provides for the following two avenues of application:

1. *Quality management* purposes, in which the facility adopts the standard as a blueprint for its internal quality system.
2. *Contractual* purposes, in which a demonstrated quality system is a condition of a contract with a customer.

Registration for Quality Management Purposes

First, companies can register and implement for quality management purposes. Companies do this to reap the benefits of getting organized and reducing mistakes. This is registration for in-house purposes only. The purpose of a quality management system is to ensure that a manufacturer's product or service (*output*) meets the customer's quality requirements. The quality system incorporates both quality assurance and quality control.

An effective quality system is methodical and procedure-driven; it is the "glue" that unites employees, equipment, and work procedures of a manufacturing site, including *suppliers* at the input end and *customers* at the output end. Facilities that operate quality management systems tend to exhibit the following attributes:

- A philosophy of prevention rather than detection
- Continuous review of the quality system
- Consistent communication within the process, and with suppliers and customers
- Good record keeping and control of critical documents
- Total quality awareness by all employees
- Management confidence

A site with these attributes will be extremely competitive.

Registration for Contractual Purposes

As said, some facilities become involved with the ISO 9000 quality system standard for contractual purposes, meaning a customer has specified an ISO 9000 quality system as a condition of awarding their contract. This condition usually requires the facility to become registered to a standard, the most common being ISO 9001. This type of registration is normally a requirement of doing business with some customers.

A facility seeks contractual registration to ISO 9001 for one or more of the following reasons:

1. It is required by customer contract.
2. Management expects contractual requirements in the near future.
3. The facility considers the registration approach as the most logical way to implement and manage a quality system.

For contractual purposes, registration requires involvement with outside agents, called **registrars**. Registrars are individuals familiar with the industry and with ISO standards that are certified with the Registration Accreditation Board (RAB) and are capable of auditing a site for registration.

ISO 9000 registration is awarded on a per-facility rather than per-firm basis. A petrochemical company with ten manufacturing sites may need to have all ten sites registered. If the petrochemical company has five process units at one site, it may want to register only one process unit. A registrar awards a firm registration after satisfactory audits of the facility's process and its documentation, and when the facility (1) has a quality system that meets the ISO 9001 standard, and (2) uses that system in its daily activities. See **Figure 4.3** for a typical certification timeline.

Figure 4.3 Typical Certification Timeline

MAINTAINING ISO REGISTRATION

The **Registration Accreditation Board (RAB)** consists of groups of accredited registrars to whom firms apply to have their manufacturing processes registered. An accredited third-party registration body (the RAB) awards initial registration after it has reviewed documentation and conducted on-site audits that verify that the firm's newly developed quality system conforms to the appropriate ISO 9000 standard. More than 75 registrars in the United States specialize in various industries.

These registrars employ auditors (certified by the RAB) to audit the processes of suppliers that have achieved registration and to ensure they are maintaining their quality system. These surveillance audits assure customers that their suppliers are continuing to conform to the ISO requirements of their manufacturing process. ISO 9001 registration is renewable and is enforced by semiannual surveillance visits by the registration body. Achieving registration is easy for firms that already have an effective quality system in place, plus once registered, registration is also relatively easy to keep. Maintaining registration through false documentation is almost impossible to fake due to the paper trail created by an established ISO quality system.

ISO and the Workforce

ISO is not just a management responsibility. When ISO registrars enter a plant to do a surveillance audit, the workers are part of that audit. Auditors can ask to see workers' training records, verify they are trained to work that area, and then ask them to perform a certain procedure exactly as the written procedure states. They can question workers about the plant's quality policy, what they do when a nonconformance occurs, and so on. In other words, auditors are verifying that the workers have been trained to be a part of the quality system.

FROM ISO 9000 TO ISO 2000

The different ISO 9000 standards are being converted to ISO 2000 standards with the intent of causing significant improvements in the quality management system. I will concentrate on the ISO 9001 standard to see how it will be affected.

The core mind-set of ISO, which had been control of the process via documented procedures, has changed to a mind-set of identifying all the minor processes and their interactions with the primary process of producing a product or service.

For instance, consider a refinery producing gasoline, fuel oils, and jet fuel. The ISO 2000 standard will look at the overall production process, which involves the following:

- inputs
- activities
- outputs
- criteria
- measurement/monitoring
- responsibility and authority

When a company became registered in the past, it had only to maintain the quality system it created. Under ISO 2000, it must show continual improvement in its system, and ensure that customer requirements are met. The company will have to show that it has a system for determining customer requirements, that the requirements are achieved, and a system for enhancing customer satisfaction is in place.

ISO 14000

The ISO 9000 and ISO 14000 families are among ISO's most widely known standards. ISO 14000 helps organizations meet their environmental challenges and is primarily concerned with environmental management. This means the organization seeks to minimize harmful effects on the environment caused by its activities and continually seeks to improve its environmental performance. Through the use of a system and standards similar to ISO 9000, ISO 14000 hopes to help companies reduce waste and environmental emissions.

SUMMARY

ISO 9000 is an ideal quality system for facilities that are serious about quality. It is also a requirement for any facility that expects to have business transactions with the European Union market and many other markets worldwide. Certification also provides a facility with a strong competitive edge. These, along with the quality system benefits previously cited, make achievement of ISO 9000 standards a powerful strategic business tool.

ISO standards contribute to making the development, manufacturing, and supply of products and services more efficient, safer, and cleaner. They make trade between countries easier and fairer. They provide governments with a technical base for health, safety, and environmental legislation. They aid in transferring technology to developing countries. ISO standards also serve to safeguard consumers and users of products and services in general. When things go well—for example, when systems, machinery, and devices work well and safely—then it is because they conform to standards. And the organization responsible for many thousands of the standards that benefit society worldwide is ISO.

REVIEW QUESTIONS

1. Briefly discuss the history of the International Standards Organization (ISO).

2. What is the purpose of ISO registration?

3. Explain the benefits of ISO registration.

4. List four documents that must be in place for ISO registration.

5. Explain why some customers would want their suppliers to be ISO registered.

6. Describe the differences between TQM and ISO.

7. Explain how continued ISO registration is enforced.

8. Describe how ISO 2000 is different from ISO 9000.

GROUP ACTIVITY

1. Divide into small groups and discuss and be prepared to answer questions. Element 4.5, Document Control. Emphasize the importance of document control in the processing industry or any industry.

Element 4.5, Document Control

Interpretation of the Standard

Document control is the control of product technical and quality system information. This does not include all documents used within the company. Documents in nontechnical departments, addressing nontechnical aspects of the product, or concerning the operation of the business need not be controlled. Many companies having implemented an ISO quality system and having discovered the benefits of document control extend their document control methods to more documents than required by the ISO 9000s, but this is not mandatory.

The intent of this clause of the standard is to maintain the body of product technical knowledge and the methodology of the quality system in a current, intact state. Technical knowledge and quality system methodology in an uncontrolled situation can be altered, diluted, or dramatically changed to the detriment of the company. The standard emphasizes the review and approval of the initial issue, and any changes to that issue are to be made by staff members capable of performing the review and approval task.

Changes to documents once reviewed and approved are to be demonstrated on the document in such a fashion as to be readily apparent to readers. It is also mandatory that readers of any document have the most current version available near their workplace, and have access to a master file for each document for the purpose of conforming currency.

To maintain this currency, the retrieval system shall ensure that obsolete documents are removed from the workplace and that no opportunity exists for documents to linger.

2. Divide into groups. Each group will write a response to the following statement:

> Standards are only important for the country you live in. You do not have to worry about standards for hair dryers, cold medication, or automobiles made in other countries.

CHAPTER 5

Total Quality Management

Learning Objectives

After completing this chapter, you will be able to:

- *Explain the methods of total quality management (TQM).*

- *Distinguish between the early concept of quality (conformance to specifications) and the new concept (maintenance of consistency).*

- *List the four major elements of TQM.*

- *Explain the importance of each of the four major elements of TQM.*

- *Explain the importance of identifying the employee as a customer.*

- *Write an explanation of why it is important to create a system for quality improvement.*

- *List examples of the cost of quality.*

- *Describe the core values of the Deming and Baldrige awards.*

INTRODUCTION

Now we continue with the evolution of quality in manufacturing and services by studying total quality management (TQM). TQM evolved from the principles of the quality gurus studied in Chapter 3. TQM is a loosely related collection of philosophies, concepts, and tools currently in use throughout the world. Because of its incongruous nature, it is not a defined program and is not practiced the same from company to company. Each company develops its own TQM program.

A TQM program requires that a company involve everything—the sum total of all its parts—in a quality management system. Some managers are tempted to involve just certain areas of their system—such as manufacturing—in a quality program. Managers may think that shipping, sales, and maintenance are not important enough to invest money and time to include them in a quality system. A total quality management system, or any quality system for that matter, doesn't work that way. It is similar to starting a physical fitness program in which you only exercise the right leg and arm. You will get some improvement in those two areas, but you have not become physically fit overall. In the same manner, a company can lose customers because its shipping department fails to fulfill orders in time, or its receiving department misplaced or sent critical materials needed in production to the wrong warehouse. It doesn't matter if the operations department has worked error-free and made on-specification product. Customers may become angry with the company and take their business to another company that has a reliable shipping department.

TOTAL QUALITY MANAGEMENT

The preceding paragraph is a good example of why companies have accepted the fact that everyone—operators, mechanics, accountants, secretaries, and vice presidents—must be trained in and understand the quality system. They are all part of the *total* system (see **Figure 5.1**). Companies involved in TQM seek to attain the following goals:

- Continuous improvement
- Defect prevention
- Customer satisfaction
- Reduced variation
- Employee involvement

Total quality management seeks quality control through prevention. A company uses quality assurance—the auditing of its quality control system—to help achieve quality. It uses statistics to create data and organize that data into useful information. The particular statistics a company uses in its statistical quality control program then define and detect nonconformances and help to improve the processes. Nonconformances are variables or

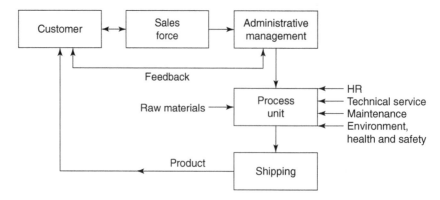

Figure 5.1 A TQM System

specifications that do not conform to the normal. TQM has moved from the engineering concept of quality, which was conformance to requirements (specifications), and now embraces maintaining consistency in products and service.

Many companies have two quality systems—ISO 9000 and TQM. If you remember, ISO 9000 is more process-oriented (though it is changing) and is not involved in safety, administration, sales, and so on. The TQM method embraces every facet of a company to improve the quality of every system in the company. Total quality management has four major components, which are revealed in **Figure 5.2.**

Customer focus is important because quality is always determined from the customer's, not the manufacturer's, point of view. It does not matter that the manufacturer and all the employees believe they have a quality product. The customers' perceptions are the only opinions that matter. Quality is always measured by the degree of customer satisfaction with a company's product and services. Because the customer's money keeps a company in business, the customer will always be king.

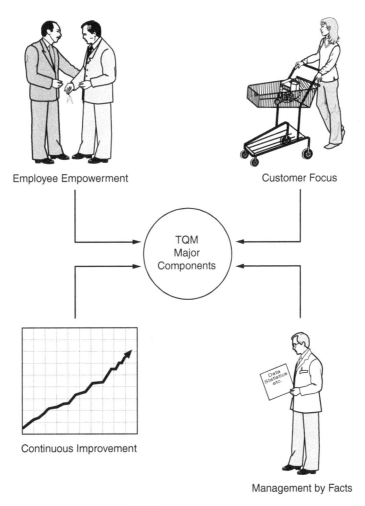

Figure 5.2   Major Components of TQM

Continuous improvement is important because improving processes will result in greater efficiency and better end products. Continuous improvement at an improvement rate of three percent is realistic and achievable. It is important to create a process that makes doing things the right way easy and doing things the wrong way difficult. Continuous improvement can also be conducted with periodical reviews of employee skills and site technology. Technology, which includes hardware, software, and people, is a vital concern of a TQM process. Technology must be reviewed periodically and updated if a company is to avoid technological obsolescence. Though continuous improvement is discussed in this chapter, Chapter 15 will revisit the subject and deal with detecting, understanding, and removing the causes of variation.

Management by data and facts is important because it brings understanding on how processes work and what improvements can be made. Managing by data and facts detects problems, weighs them by importance, and then aids in planning the elimination of the problem. Sometimes the data and facts do not reveal a problem but do reveal wasted man-hours, materials, utilities, and so on. Managing by data and facts is covered in Chapters 13, 14, and 17, all of which address quality tools, statistics, measurement, and statistical process control.

And finally, **empowering employees** is important because management must get the buy-in and commitment from its workers. The workers are the people on the process unit or shop floor who know how the system really works. They have the knowledge and experience to detect and help eliminate nonconformances. Management can empower employees by giving them the right to question, challenge, and attempt changes in the way products are produced and services are rendered. Empowering employees is covered in detail in Chapter 6.

Failure of a Quality Program

Many quality programs and initiatives have been created in the last thirty years, and many new ones will crop up in the future. Numerous firms may have tried at least three of the quality programs listed in **Table 5.1.**

Table 5.1   Quality Initiatives

Quality circles
Work teams
Employee suggestion programs
Participative management teams
SQC (statistical quality control)
Reengineering
Job-enrichment task forces
Performance excellence
Six sigma
Flattened organizations
Strategic business units

Companies that have undertaken and then abandoned two or more of these programs have not approached the quality issue correctly. Quality problems cannot be approached with just emotions and gritted teeth the way soldiers charge enemy machine guns. They must be attacked head-on and logically by installing a total quality management (TQM) system. Tools, incentives, and rewards must be used to encourage workers to use the program, and management must show its commitment to the program.

Management's Role in TQM

Management has the most critical role in TQM because management has the power to dictate policy, as well as the money to implement what is needed. Management has a constant juggling act with production, safety, and quality. Management must determine how much money and attention can be focused on each. It is not an easy task, and someone is always complaining that management sacrifices one of the three. Management must seek to avoid the following six serious issues that are considered root causes of quality problems:

1. Placing budgetary considerations ahead of quality
2. Placing production scheduling considerations ahead of quality
3. Placing personal politics ahead of quality
4. Lacking fundamental knowledge of the site processes
5. Believing their position entitles them to special treatment or exceptions
6. Practicing autocratic behavior that de-motivates employees

The majority of managers avoid these mistakes though they often struggle with the first two issues because they never have enough money to do everything that needs to be done, and production makes the product that is sold and brings in the money to keep the process unit running.

BARRIERS TO TQM

What barriers keep people from embracing TQM and working in the new way of continuous improvement? A barrier is any policy, procedure, reward, punishment, attitude, and/or method of personal interaction that inhibits people from thinking, talking, working, and acting in support of the new management system. The following major barriers appear in many businesses:

- Lack of commitment by top management
- Concerns about the ways people are judged—performance appraisals or merit ratings
- Poor communication or poor treatment
- Employment security concerns
- Inadequate training
- Resistance by unions due to the threat of lower employment levels

The order of importance, of course, varies by organization. At some companies, concern about employment security is the most serious barrier, whereas other companies have major problems with their performance review and pay systems. But the one that usually heads the list is lack of commitment by top management.

Table 5.2 Management Philosophy Systems

Authoritarian (Pre-1980s)	Participative (After 1980)
Fear of reprisal	Open atmosphere
Position represents knowledge	Workers are the experts
Status quo is good enough	Continuous improvement
Conformance to specifications	Quality defined by customers
Driven by management and cost	Driven by external customers
Individuals rewarded	Teams rewarded
Productivity versus quality	Productivity through quality
Driven by financial performance	Driven by quality measures
Short-term perspective	Long-term perspective

TQM should drive out fear and open up lines of communication. The TQM system must show people that they are recognized, rewarded, and promoted on the basis of their behavior to the quality system. It should replace the old authoritative control system with a participative system (see **Table 5.2**). It should ensure that people receive the training they need to work successfully in the new quality system. It should help unions understand that waste is the most serious cause of employment insecurity. As a company begins to work in the new way, employees typically react by believing that nothing is happening fast enough, that leadership is lacking, or that training is inadequate. Do not be discouraged. A major culture change does not happen overnight; it may take several years and rarely goes smoothly. Everybody, including management, needs extensive training on the new system. Many people may resist and and wait to see if the changes will go away and be replaced by some other flavor-of-the-month program.

The organization should change gradually to eliminate barriers to continuous improvement as these barriers become apparent. The organization can, however, begin immediately promoting people who are working in the new way. This approach has the advantage of showing people a direct causal relationship with the new way of working. Management can constantly look for opportunities to make visible changes, which then serve as both proof and reminder of their commitment to working in the new quality system. The goal is to have people become part of the team, to want to share in the success of the organization, and to recognize that only through the success of the organization can they themselves be secure and successful.

INITIATING TQM

TQM has a universal appeal because it is a long-term system to achieve customer satisfaction through the elimination of waste and the continuous improvement of a firm's products and services. When initiating TQM, management must avoid creating the impression of, "Here comes another program with a coffee mug and poster that no one will remember six months from now." One of the best ways to gain employee enthusiasm for starting a TQM system is through viewing employees as the internal customers.

The Employee as Customer

Meeting customers' requirements means listening to the customers, responding to what they want, and doing what is agreed upon. Ishikawa defined internal customers as the people within the company who receive the work of another and then add their contributions to the product or service before passing it on to someone else. In manufacturing, the internal customer is the next person down the assembly line. A company's processes could have quite a few internal customers. At one well-known shoe manufacturer, a pair of shoes passes through a hundred pairs of hands from start to finished product, boxed and sitting on the loading dock. That's how far the chain of internal customers stretches around that factory. In a restaurant, the chef has assistant chefs, waitstaff, and a *maître d'* as internal customers, and the chef must meet their requirements if the diners are to experience the chef's part in customer satisfaction—a delicious meal graciously attended at a fair price.

Many companies look upon their employees as Ishikawa's internal customers because creating a philosophy and culture of customer service throughout the company enhances production, and ultimately, external customer satisfaction. All company employees should view everyone that receives a service or product they produce as customers. Just as a company wants satisfied external customers, it also wants satisfied internal customers. If the operations department makes on-specification product, if all departments work efficiently, and if interdepartmental liaisons and functions work seamlessly, then the company will likely also (1) minimize waste (which increases profits), and (2) satisfy the external customers (which will retain those external customers).

Examples of poor customer service abound. We have all experienced it in restaurants, grocery stores, department stores, or phone service providers, and so on. In a processing plant, the operator may experience poor customer service from maintenance, the quality control laboratory, technical support, or human resources. Poor internal customer service can result because it is simply easy to forget the importance of satisfaction when handling everyday problems with coworkers. For instance, if an operator brought a pump down for maintenance work and maintenance arrived two hours late, then the operator had to block-in the pump, clear it, do a lockout/tagout, and have his supervisor verify everything. If an operator has to use 16-ounce sample containers because the warehouse is out of 4-ounce bottles, the operator will have to carry two sample baskets instead of one, carry the extra sample weight, and collect 12 ounces of extra sample material, most of which will be discarded by the quality control laboratory and wind up in a slop tank.

Small everyday problems don't seem like big issues but they are because they impede efficiency, allow waste to continue, and prevent the process of continuous improvement. Small everyday problems waste time and resources and can lead to bad feelings that prevent smooth operations. These problems accustom us to accepting imperfect systems instead of eliminating the imperfections. Training programs can educate us to the importance of internal customers, but only an ongoing quality improvement system will reinforce the training and sustain the quality initiative.

Poor Treatment of Internal Customers

Involving the entire company in the process of customer service can reinforce the importance of providing quality services. Not all support departments or shared services systematically attempt to identify their users' needs, even though employees are internal customers of those support departments. When all the departments within a process (making shoes, gasoline, chemicals, and so on) work toward the same goal, the process is optimized. The process is less than optimal when some departments work toward their own goals without regard for the rest of the process. Customer satisfaction, unfortunately, is not always directed toward internal customers. Internal customers might be treated poorly for the following reasons:

- Internal competition
- Short-term thinking
- Lack of knowledge of internal customer requirements

This last item is the principal reason why internal customers are treated poorly, and is an example of the left hand not knowing what the right hand is doing. Only members of support departments can change this by meeting with members of other departments and receiving firsthand knowledge of how they are failing the internal customers. With these data, those in the support department can incorporate changes in their actions that support or enhance the departments that depend on them.

TQM FOR PROFIT

A great deal of profit can be made by quality improvements in products, services, business processes, and people. In 1986, internal analysts at IBM put the cost of nonconformance or failure to meet quality standards in its products and services at a minimum of 11 percent of revenue, or $5.6 billion. Much of this loss was due to the costs of having poor business processes.

Management must create a system that prevents defects from happening in the company's various processes. To accomplish this, a company has to act now on situations that may cause problems later. The idea of prevention—stopping things from going wrong—is pivotal to TQM programs. Why should people spend time sorting out complicated quality problems when they could have been prevented in the first place? Yet, although most people believe in prevention, it remains a hard struggle or simply unattainable for most. *To err is human* becomes a universal escape clause for poor performance. *Do it right the first time* and *zero defects* are great sound bites, but they are **performance standards** that are hard to accept in conventional work practices that seem to have an anticipation of failure built into them.

The 1-10-100 Rule

Catching and fixing problems before they leave the manufacturing site greatly affects profits. The 1-10-100 Rule, which business studies have shown applies to many activities, shows just how much money can be saved (see **Figure 5.3**). The 1 in the rule represents one dollar, the 10 represents ten dollars, and the 100 represents 100 dollars. Figure 5.3 illustrates how the further a nonconformance moves through the system, the more expensive fixing the problem becomes.

For example, if an employee working on an assembly line does something wrong but catches it before it leaves the area, the expense is minimal expense (a few minutes are wasted). If an incorrectly assembled object moves 100 yards down the line and someone notices it, the line might have to be stopped and parts and people run to fix the defects. This is greater waste and more expensive. However, if employees load a nonconforming product

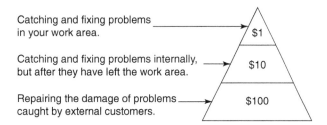

Catching and fixing problems in your work area. — $1

Catching and fixing problems internally, but after they have left the work area. → $10

Repairing the damage of problems caught by external customers. → $100

Figure 5.3 The 1-10-100 Rule

on a truck that leaves the plant, the problem has suddenly become very expensive when a customer attempts to use the defective product that doesn't work. The losses incurred include the costs of (1) losing that customer, (2) paying for all damages to the customer's process, (3) sending a repair team to the site, or (4) bringing a truckload of defective material back to the site. This can become very expensive.

TQM and Teamwork

Total quality management cannot work without teamwork. This topic will be discussed in detail in Chapter 8; however, a few things will be discussed to complement this chapter. A company or corporation must be thought of as a huge team composed of hundreds, if not thousands, of team players. They all must work together to win the game—the "game" being prospering another year and gaining market share.

TQM has no chance of success if team members do not cooperate. Every team member is critical, from the CEO behind a mahogany desk to the operator on a process unit. Think of all the teamwork involved in a professional football team. The management team determines how much money they have available for players' and coaches' contracts, the public relations team, the scout team seeking replacement players, the coaching team (special teams, quarterback, defense, offense), cheerleaders, sales promotions, and so on. They all have necessary tasks to perform for the spirit and survival of the team.

ACCEPTABLE QUALITY LEVELS

The goal of any quality system is zero defects. A company cannot afford to have a dual performance standard—a standard that says we want zero defects but we know achieving that is impossible so do the best you can. This leads to a willingness to put up with acceptable quality levels at work and a system that is defective. This is best illustrated by a real case experienced by an IBM firm in Windsor, Ontario. The company ordered a shipment of components from a Japanese firm, specifying the acceptable quality limit of three defective components out of every 10,000. In a cover letter that accompanied the fulfilled order from the Japanese supplier to the IBM firm, the Japanese company explained how difficult it was to produce the defective parts, and said: 'We Japanese have a hard time understanding North American business practices. But the three defective parts have been included and are wrapped separately. Hope this pleases." The Japanese company gave IBM what it specified—three defective parts and 9,997 good parts!

Acceptable Quality Levels Away from Work

Acceptable quality levels (AQLs), an industry concept since the end of World War II, offer a diametrically opposed mind-set to total quality. Instead of getting it right the first time, the company encourages defects by setting AQLs. Do you think 99 percent on-specification is a good quality level? If Americans tolerated an acceptable quality level of 99 percent, the results would be the following:

- At least 200,000 wrong drug prescriptions each year
- More than 130,000 newborn babies accidentally dropped by doctors or nurses annually
- Unsafe drinking water almost four days each year
- No electricity, water, or heat for about fifteen minutes each day
- Undelivered newspapers four times each year

Would you willingly settle for any of these 99 percent acceptable levels? Probably not. Think about acceptable quality levels in your personal life. Would you accept being shortchanged once a week when you went shopping? A surgeon performing an operation on your wife and removing the wrong organ? A checker at a grocery store forgetting to bag the T-bone steaks you bought? How about acceptable quality levels for open-heart surgery? Pacemakers? Automotive air bags? The goal is Crosby's zero defects. Nothing else is acceptable.

Set the same standard where you work. You don't have as much control there, but you can gradually influence changes that lead toward zero defects. Remember, zero defects is an attitude, a state of mind that does not tolerate defects in materials or services and seeks to eliminate defects.

Acceptable Quality Levels at Work

Assume you have been hired as an operator in a refinery. The supervisor of your processing unit calls you in and says, "Listen, we want zero defects here but we're all human. You are going to make some mistakes; we all do. Just don't make any really bad mistakes, okay?"

Would you feel comfortable making a mistake in a refinery where highly flammable materials are under high temperatures and pressures? Would you feel comfortable knowing that everyone in the four five-man crews on your unit got the same message—that making a mistake is okay? Are you being assigned to a unit with a culture of sloppiness and accidents? Do you feel relieved that you and your coworkers are *expected* to make mistakes?

Contrast this scenario with another one. Assume you have been hired as an operator in a refinery. The supervisor of your processing unit calls you in and says, "On this unit, we believe in working safe and being safe. We do not tolerate mistakes. We investigate them all, and we eliminate their causes. If you make a mistake, don't try to hide it. Tell us; we'll train you so that you don't make that mistake again. We do not expect you to make the same mistake twice. Do you understand?"

With which supervisor and unit culture would you like to work? The answer is obvious—the unit with the safest work environment. The unit is not perfect, but it has the zero defect attitude.

PERFORMANCE STANDARDS

No TQM system will perform satisfactorily without performance standards. Managers and management systems account for 80 percent of what is commonly called quality problems. Once management is on board the quality train and actively encouraging continuous improvement and prevention for all processes, it must establish an effective performance system that includes every individual in the organization. A performance system measures how well a person or process unit performs to defined standards.

All companies and processes have performance systems and standards. A performance system will ensure that:

- Quality expectations (standards) are established and understood.
- Barriers to quality performance are removed.
- Regular, specific feedback on quality is provided.
- Rewards are provided for quality performance.

A performance standard is management's way of telling employees how often they want employees to do things right. Some employees might have the attitude that doing things right all the time is impossible, or that they are working at acceptable quality levels. However, these same workers in their personal business demand zero defects. They want their cars' brakes fixed correctly, and they expect electronic appliances to work as described. The same performance standard should be expected at work, because individuals won't take the necessary actions to improve their performance or the system's performance if they believe they have the option of not doing things right the first time. Over time, this institutionalizes the acceptability of error.

Some typical performance standards in the processing industry include the following:

- How many times you can be tardy in a six-month period
- How many times you can be absent in a six-month period
- How many times you can make the same mistake with equipment
- How many times you forgot to do something you were required to do

If an employee works in a system that is defective no matter how judicious and sincere the worker is, he is bound to experience "bad" days. It is not his fault (Remember Deming's red bead experiment?); rather, it is the fault of the defective system. The employee did not fail; the system failed. However, the employee may be blamed for the failure. This will affect attitudes toward work, which, in turn, will affect productivity and efficiency. Management is responsible for providing equipment that works correctly, training to upgrade skills, and materials as needed, and for creating a decent work environment—not too hot, not too cold, not too cluttered.

When a business has a good performance system in place, employees will have access to environmentally good work conditions, sufficient skills, and logical work systems to help them do what they are inclined to do anyway—take pride in their work and produce high-quality products and services.

QUALITY AWARDS

The Deming and Baldrige awards have been included at the end of this chapter, because like TQM, they are methodologies with which to build quality systems. Japan encouraged the pursuit of quality in its goods and services by creating the Deming Award, named after Edwards Deming, the American who went to Japan after the end of WWII and helped rebuild its shattered industries. To encourage American businesses to pursue quality in the

United States, the Reagan administration initiated the Baldrige Award, named after former Secretary of Commerce Malcolm Baldrige who died in 1987. Both awards are highly regarded by government and industry.

The Deming Prize

The Japanese Union of Scientists and Engineers (JUSE) created the Deming Prize in 1951. A main focus of the prize is efficient statistical use of data. The award has the following two main categories:

1. Deming Prize for individuals
2. Deming Application Prize (DAP) for organizations

The Deming Prize has the following ten main assessment criteria:

1. Strategy
2. Organization and management
3. Education
4. Information
5. Analysis
6. Systems
7. Quality control
8. Continuity
9. Results
10. Action and reaction

The Malcolm Baldrige Award

The Malcolm Baldrige Award was founded in the United States in 1987, to promote total quality management as an important method of improving American products and services. The award recognizes large- and small-service and manufacturing companies that demonstrate exemplary quality in their products and services. The Baldrige Award provides companies with a set of standards against which they can evaluate their organizations. Most U.S. companies are familiar with the Baldrige Award, and many use its criteria to develop their own approach to total quality. The criteria have become a blueprint for how to run a better company. The following are core values built into the Baldrige criteria:

- Customer-driven quality
- Full participation
- Management by fact
- Continuous improvement
- Design quality/prevention
- Fast response
- Long-range outlook
- Partnership development
- Public responsibility

SUMMARY

The heart of TQM is customer focus. If an organization wants to meet customer needs— both internal and external—it must have the proper equipment, procedures, and systems. Depending on the type of complaint and product, swiftly responding to a customer complaint might entail availability of analytical services, technical personnel who can analyze the problem, and quick and effective means of communication. For multinational companies that do business all over the world, fast and effective communication is a must. The following are major components of TQM:

- Continuous improvement
- Management by data and facts
- Employee empowerment and involvement
- Customer focus

Experts also believe that TQM requires companies to review employee qualifications, promote technical skills development, and instill the spirit of teamwork. Acceptable quality levels in industry can be misleading. The only acceptable quality level is zero defects.

Teamwork is critical to TQM. A company is a huge team composed of multitudes of small teams, each with tasks to perform that ultimately result in customer satisfaction and profits.

REVIEW QUESTIONS

1. Describe the philosophy of total quality management (TQM).

2. Explain the methods of total quality management.

3. Distinguish between the early concept of quality (conformance to specifications) versus the new concept (maintenance of consistency).

4. List the four major elements of TQM.

5. List two major barriers to a successful TQM program.

6. Explain the fallacy of having acceptable quality standards.

7. We all know that a 90 or better is an "A," which is the ultimate grade. You are about to have a pacemaker installed inside your body. The company that makes the pacemaker rates their product as 99.6 percent error-free. Describe your feeling the night before the surgery as you contemplate the defect rate of this product.

8. For problem #7, how many out of 1,000 pacemakers will be defective?

9. Explain the importance of each of the four major elements of TQM.

10. Define the following terms:

   External customer

   Internal customer

   Customer satisfaction

11. Explain the importance of identifying the employee as a customer.

12. Describe the 1-10-100 Rule.

13. What is meant by the term "performance standards"?

14. Why are performance standards important?

15. List the core values of the Baldrige Award.

16. List the asset criteria of the Deming Prize.

GROUP ACTIVITIES

Divide into small groups and choose one topic below to discuss and then report your conclusions to the class.

1. Make a list of experiences you had as an employee in your previous jobs when you were not treated as a customer, and explain how this affected your morale and attitude.

2. Using that same topic, explain how those experiences affected you or your company's productivity and efficiency.

3. Read and discuss the following article:

A Lotte of Ambition

Article by Mike Speegle, August 2008

Women in South Korea have the reputation for being some of the world's most demanding shoppers. Keen bargain hunters have an eye for quality and, if something fails their expectations, they are not shy of venting their anger. They photograph shops or goods that fail to make the grade and post the pictures on the Internet with detailed and withering criticism. The offending business may then be shunned by other buyers. So exacting are the demands of Korean customers that Western firms often solicit their opinions of new products before launching them in Europe or America. If you can please a Korean customer, you can please almost anyone.

Accordingly, South Korean department stores provide some of the best service in the world. With a smile and a bow, sales assistants scurry after customers, attending to their whims while calming their upset children. Lotte, South Korea's biggest department store chain, thinks its experience at home makes it ideally suited to serving the new rich in other fast-developing economies. In July 2008, Lotte plans to open a huge department store in China in the Wangfujing shopping district in Beijing. It has set an annual sales target for the store of $150 million. The wealthiest customers will be granted special parking spots and guided around the store by personal attendants. Appealing to the very rich works well for Lotte at home: its richest 1 percent of customers accounted for 17 percent of its $5.8 billion in sales last year. Lotte's Beijing store will be a test of the company's plans to launch nine more stores in other big Chinese cities. Lotte has asked consultants to analyze the Chinese market and help it choose a combination of foreign and local brands to appeal to Chinese shoppers. Lotte plans to sell South Korean cosmetics and clothes, counting on the appeal of Korean pop culture, which is popular throughout Asia. Lotte's Beijing staff has been sent to Seoul to learn about its procedures, marketing, and service.

It hopes China will not prove such a tough nut to crack. "Many retailers find China tough," said Lotte's director of international business development, "but we're quite comfortable with the prospect."

CHAPTER 6

Customer Satisfaction

Learning Objectives

After completing this chapter, you should be able to:

- *Write a definition of a customer.*

- *List six ways to measure customer satisfaction.*

- *List four variables used in customer satisfaction surveys.*

- *Describe two things a company can do to keep the customer satisfied.*

- *Explain why maintaining customer satisfaction is more difficult as a company increases in size and expands its locations.*

- *Explain why companies use training to maintain customer satisfaction.*

- *Explain the importance of going beyond customer satisfaction.*

INTRODUCTION

"Who is the customer? The customer is the most important visitor on our premises. He is not dependent on us. We are dependent on him. He is not an interruption of our work; he is the purpose of it. He is not an outsider in our business, he is part of it. We are not doing him a favor by servicing him. He is doing us a favor by giving us an opportunity to do so." This is a quote by Mahatma Gandhi, critic of Indian officialdom. His words are displayed in public offices all over India.

You are probably asking yourself what does customer satisfaction have to do with quality? If you think about what you've learned so far, customer satisfaction is a principal concern of the quality process. **External customers** are the people keeping companies in business by buying their products or services. Business owners want customers to buy from them, keep buying from them, and tell all their friends and relatives to buy from them. They do not want to experience a surplus of unsold products at the end of the year, which Chrysler experienced in 2007, when it was unable to sell thousands of cars.

Business owners should never become complacent enough to assume they know what customers want. Many companies make errors in this area, introducing products such as the "new Coke" and the Ford Edsel that didn't sell. Organizations expend considerable time, money, and effort determining the "voice" of the customer, using tools such as customer surveys, focus groups, and polling.

THE CUSTOMER

Customers of all businesses have both specifications and expectations. These are not identical. **Customer specifications** are quantitative descriptions used to define the quality of a product or service. Specifications are always included in contracts and purchase orders between customers and suppliers. An example is the weight percent titanium oxide used in a premium brand of paint or a minimum of 97 percent on time delivery service. **Customer expectations** are qualitative factors, such as clean restrooms, floors, and tables in a restaurant. Another example is expecting to be waited upon immediately and with pleasant service attitudes. Customer expectations are a product or service you expect wherever you do business.

Meeting customer expectations is often complicated because customers don't always express their requirements. For most products and services, a mixture of stated and unstated customer requirements is normal (see **Figure 6.1**). Stated requirements are those that customers express when making the commitment to buy. For example, when booking a hotel room, a customer might specify that a king-size bed in a nonsmoking room is required. The customer may not have specified that a high-speed Internet connection is needed to connect with her company's network. Assuming that this will be available, the customer will be disappointed to discover that the room has no high-speed connection, and that the only place to plug a computer into a phone line connection is in the hotel lobby.

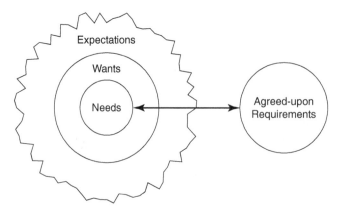

Figure 6.1   Clarify Expectations

Customers usually want outcomes more than they want a specific product. For example, assume a young man with appendicitis is admitted to a hospital. The cure alone is his desired outcome. On the other hand, a cure might not be the best outcome. Perhaps the best possible outcome would be that his appendix is removed, he's cured in a few days, and he meets a lovely, rich young lady while in the hospital. They fall in love and get married. Contrast this outcome with an outcome where the young man is cured of appendicitis but, while in the hospital, contracts a blood disease and dies. The hospital can't guarantee every young man a rich wife with each stay, but it can identify possible poor outcomes and work to eliminate their potential causes. The presence of a telephone data port connection in the hotel room and sanitary conditions to prevent the transmission of blood diseases in the hospital were never specified by the customers but were required (expected) anyway. As the examples illustrate, these situations are much more common than we might at first suppose. So how much effort should a business expend trying to understand what the real customer requirements (plus expectations) are? As much as it takes.

What is Customer Satisfaction?

Without satisfied customers, a company is going out of business—unless the company is a monopoly. Customer satisfaction is the principal concern of the quality process. Realizing that, you might wonder:

- What is customer satisfaction?
- How do you maximize and maintain customer satisfaction?
- Can a company grow too large to maintain customer satisfaction?
- Does the expense of customer satisfaction reduce profit margins?

These are important questions, and the answers to them will help establish a solid understanding of several concepts used to achieve quality.

You may be thinking that customer satisfaction is really nothing more than satisfied customers, and you are absolutely correct. *Satisfied customers* mean *customer satisfaction*. It's that simple, but not simple at all. If you own a small company in a small town, many people may know you, and they may like to buy from you based on your integrity, service, and product quality. The fact that your customers keep coming back is a good measure of customer satisfaction.

Importance of Satisfied Customers

There are no important results internal to a business. The most important results of any business are satisfied external customers. The results of a hospital are healed patients. The results of a school are pupils who have learned and apply their knowledge. Inside a business, there are only cost centers. Results exist only on the outside of businesses—in other words, with the customers.

Satisfied customers are important because today there just aren't enough customers to go around. We can produce more washing machines, automobiles, clothes, and TV sets than can ever be sold. Customers are perceived as scarce and precious as more countries (China, India, Taiwan, Malaysia, Turkey, and so on) develop and expand their manufacturing bases and compete for customers. Just as various manufacturers produce TV sets, various manufacturers also produce styrene, paraxylene, and other commodity chemicals in the processing industry. The nations of the world are like stores at a shopping mall, and the customer picks which store has the best quality at the best price.

Regardless of what business companies are in—manufacturing, retail, or service—they have customers. Because customers have free will and can shop anywhere, creating a pleasing customer experience is critical to keeping customers. If a family spends $100 a week at a grocery store for the next 20 years, that family is worth $104,000 to that grocery store. Losing just ten customers can mean losing more than one million dollars. Companies can no longer compete on price or location alone. How many times have we seen a Lowe's or Home Depot, or an AutoZone or O'Reilly's in the same shopping center offering merchandise at almost identical prices? A pleasing customer experience is a requirement for staying in business.

WHAT IS A CUSTOMER?

Merriam-Webster's Collegiate Dictionary defines a **customer** as "one that purchases a commodity or service." However, some people consider that definition too imprecise. Consider external customers. A person who buys a new car is clearly the customer. A man buying groceries is clearly a customer. However, is the person buying the new car the only customer? What about that person's family, or others who might travel in the car? Is it important that they be pleased by the operation of the new car? Consider the man buying groceries. Are there still other customers (the people he will cook for)? Manufacturers of branded consumer goods always have two customers at the very least: the grocer who stocks the shelves and the housewife who empties the shelves. Are the car and grocery buyers the primary customers and the other people that contact the car or groceries secondary customers? It can become a complicated situation.

DISSATISFIED CUSTOMERS

Satisfying customers is a simple concept. It involves defining and then meeting customers' needs and expectations. Fully developing those needs can be as complex as performing a quality function deployment, or as simple as listening to customers. Satisfying customers has to be a paramount concern, as dissatisfied customers represent major lost opportunities.

The cost of a dissatisfied customer extends well beyond the business lost of that one customer. Typically, 95 percent of service time is spent addressing external customer problems; only 5 percent is spent on prevention. For the reasons that follow, it is important to have employees well trained in the importance of customer satisfaction. The following are some startling facts.

- Recent studies confirm that developing new customers to replace dissatisfied customers costs an average of five times more than retaining satisfied customers does.
- To make up for one disappointing incident, customers must have twelve positive experiences.
- The average dissatisfied customers tell nine to twelve other people about poor service or product quality.
- By making it easy for customers to voice their complaints, listening and responding immediately to those complaints, and treating customers with respect, 74 percent of dissatisfied customers can be won back.

Dissatisfied customers have more than one way to get even with companies that fail in their responsibilities of service and product, especially in the Internet age. The computer keyboard has become a mighty sword. Thanks to the Internet, angry customers can get even while enjoying the company of fellow sufferers. *Business Week* calls these tech-adept complainers "consumer vigilantes." As a true example, a high-speed Internet provider (call it HighSpeed) antagonized a customer who decided to become a warrior for the consumer. He launched a website called HighSpeedMustDie.com, where similarly frustrated consumers gathered and wrote up their stories. Before long, HighSpeed learned of the website and began contacting each angry customer, offering to fix their problems. The lesson? HighSpeed should have paid attention to common sense: angry customers are bad for business. Now the company is doing what it should have done in the first place but *after* angry customers broadcast their mistreatment on the Internet to an untold number of potential customers.

MAINTAINING CUSTOMER SATISFACTION

Organizations cannot afford to be passive when the subject is defining sources of customer dissatisfaction. A strong proactive stance is required. No one can assume that the customer is satisfied just because no complaints are received. Research in this area shows that most dissatisfied customers will not complain to the organization or individual with whom they are dissatisfied. The research also shows that nearly all dissatisfied customers will never return with additional business, and they will tell other potential customers about their dissatisfaction.

Surveying to Maintain Customer Satisfaction

As a company grows, direct access to its many customers becomes almost impossible. For large companies who want to maintain customer satisfaction, surveys have proven to be very useful for identifying customer feelings and expectations. However, on many surveys, the wrong people are surveyed, or the wrong questions are asked, are asked in the wrong way, or are asked at the wrong time. Businesses must measure customer satisfaction (or dissatisfaction) if they intend to remain in business, but the survey must be conducted correctly and must reveal valid information.

Repeat customers and surveys are not the only ways to measure customer satisfaction. The following are some other ways:

- Letters of appreciation from customers that are generally not solicited by the company
- New customers obtained by references or "word of mouth"
- Absence of or low rates of customer complaints (although many disappointed customers do not write complaint letters)
- Brand loyalty or name loyalty
- Increasing business (an indicator that may not give a direct true measure of customer loyalty)

In essence, these all are commonsense methods of measuring customer satisfaction that give a glimpse at the quality customers perceive in products or services. However, when a company compares its customer satisfaction with a competitor's, it may be seeking to understand minor differences in customer satisfaction. Many market researchers assign values to each variable affecting customer satisfaction, such as the following:

- Wait time of the service
- Courteousness of the service
- The availability of the object the customer seeks
- Correct billing

To come up with the indexes, market researchers devise a system that gives weight to each variable that can affect (or that are believed to affect) customer satisfaction, depending on its importance to the business. For instance, repeat customers may be a category that carries a larger weight than the price of an item. Generally, unless the price is extraordinary, the price is not that big a factor in customer decisions. Quality and reliability are more important. Similarly, customer complaints may carry a larger weight than returned merchandise. When accepting returned merchandise, some companies try to gather information about the real source of the customer's disappointment with the product. As competition intensifies, these types of surveys have become more and more important.

Customer surveys are valuable in the sense that they help vendors gain an understanding into customers' buying habits. Whether or not we realize it, we are participating in surveys when we use our supermarket or pharmacy cards, which give us significant discounts on some products. When the card is scanned, our names and what we have purchased go into a database. This is a survey of our buying habits. What do surveyors do with the index numbers that yield a quantitative picture about the company's sales and sales distribution? With the index, a company can determine the dollar value of each point on the index. A point on the index may be worth several million dollars for large companies like Exxon, Target, and General Electric. A point drop in the customer index could translate into a loss of several million dollars. Taken the other way, a point increase in the index could translate into an increase of millions of dollars in revenue.

Although many indexes are linear—that is each number on the index has the same value—many other more sophisticated indexes may not be linear. Many psychologists feel that an index of human satisfaction can't be linear. They assign the concept called "diminishing value," meaning that the high levels of customer satisfaction may not be that different from each other. Many psychologists doubt the scientific basis of such indexes. They claim that to measure human satisfaction by numbers is too simplistic and should not be attempted at all. Many business executives, however, find such linear indexes very useful for making rational decisions.

Keeping the Customer Satisfied

Making and keeping customers happy are the two rules that guide a company into successful longevity. Most companies hope to be in business for centuries, if not forever. DuPont was founded in 1802 (over 200 years ago); Ford Motor Company founded in 1903 by Henry Ford with just $28,964 dollars (over 100 years); and Deere and Company was founded in 1837 (over 170 years ago). The average life expectancy of a successful business today is 30 years. To run a business successfully, companies need to keep their customers happy. This is easy to say but not easy to do. So, how does a company keep customers satisfied? The following are a few things they can do:

1. Keep prices reasonable. This does not mean that the company keeps the prices at the lowest levels. For example, if you like Shell gasoline, and its unleaded sells for $2.30 per

gallon and the competitor brands sell for $2.28 a gallon, you will still buy Shell gasoline because the difference is not significant. However, if Shell raises its price to $2.75 per gallon, it may lose customer loyalty because the price difference is too great and creates a perception of greed. But, for some products, there is this counterpoint. In a commercial for L'Oreal, an expensive skin care product line for women, the company advertises that you can buy competing products for one-fourth the price. L'Oreal's message is that its product is not cheap, and the lady (in the commercial) says, "I am worth it."

2. Give improved service. The word *service* means different things to different businesses. For instance, in the cafeteria business, service may mean how quickly the business delivers your food. Even in fast food restaurants, how quickly customers receive the food is important. In fact, McDonalds built their prestige by offering quick delivery of their food. In department stores, service may refer to how clerks help customers find sizes that aren't on the racks or check out faster.

3. Make the purchasing experience pleasant for the customer. How many times have you wondered why people pay a higher price to go to clean restaurants with a special ambience (with an ocean overlook, candles, and small orchestra)? Similarly, several chains of grocery stores that do not have the lowest prices provide extra shopping dividends like clean stores, well-organized shelves, and quick checkout services. One furniture outlet in the Houston area feeds its customers, entertains their children while customers shop, guarantees next-day delivery, and assists with financing. These values (marks of quality) create customer loyalty.

4. Survey the changing world of customer needs and desires. Remember the single most important word of the business world is *customer*. Like anything else, customers' needs change over time, and businesses must change to meet those needs (see **Figure 6.2**). One example of a company with a sizzling product that failed to keep up with changing expectations was Atari, one of the first makers of computers. In the early 1980s, Atari computers were very popular and simple to use. As the years went by, other powerful computers with better screens came on the market. Atari didn't change in time to satisfy customer needs, and its product became extinct. Cell phones are also a fiercely competitive business. Each year they are made smaller, more stylish, and have more features (phone, e-mail, camera, PDA, and so on). Cell phone manufacturers are constantly seeking to provide innovations the customers will desire.

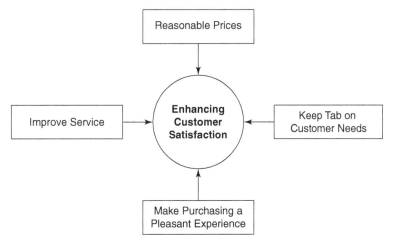

Figure 6.2   Tools for Gaining Market Share

BEYOND CUSTOMER SATISFACTION

W. Edwards Deming was onto something when he said that the customer's definition of quality is the only one that matters. As a business owner, you know what quality means to your company, but do you know what it means to your customers? Customers can tell you what they value about your core products and the surrounding support services that make up their experience. Customers who are merely satisfied are easy for your competitors to entice away, or they may leave just for a desire for change. To retain customers, some companies go beyond customer satisfaction.

Companies could ask two simple questions on a customer survey and receive a lot of valuable information. The two questions are: (1) How satisfied are you with our product and/or services? and (2) Would you recommend our products and/or services to a friend? The first question addresses basic satisfaction; the second question addresses true commitment. See **Figure 6.3**, which is a simple diagram of satisfaction versus commitment. Note that satisfaction falls only in the middle range of the continuum. The blunt reality is that basic customer satisfaction is no longer adequate for businesses to remain successful. Basic satisfaction just means that customers might continue to use your products or services unless a better offer shows up. Satisfaction is little more than the absence of dissatisfaction. The ultimate goal for a business is committed customers who become advocates.

To succeed, companies must strive to turn satisfied customers into loyal customers, and turn loyal customers into advocates who tell everyone how great the company is and that they would never think of buying from anyone else. These are the top rungs of the customer experience path (see **Figure 6.4**). Even before prospects become customers, business owners can start informing customer expectations. Potential customers see commercials or hear about the performance from their friends. Once they become customers, a business owner's goal is to deliver what was promised and ensure that the customers are satisfied. Beyond satisfaction, owners must strive to ensure that they deliver consistently positive experiences and build a strong relationship that develops loyal customers, and ultimately, advocates.

Loyal customers are those who intend to stay with a company. Customers will overlook occasional failings when a company builds loyalty. Even though outcomes don't always go as planned, loyal customers believe that the company strives to do its best and will cut a little slack to resolve a problem.

Advocates are even more valuable to a company. ***Advocates*** are customers who feel a strong connection to a product or service and will actively endorse it to others. They are so impressed with a company that they are willing to stake their own reputation on endorsing

Figure 6.3 The Quality Continuum

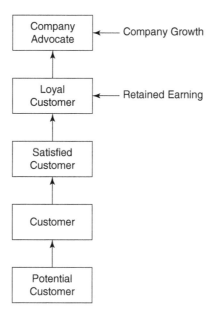

Figure 6.4 Customer Experience Path

the product to their friends. Advocacy is the best predictive indicator that an organization is delivering product and service experiences that are important to its customers. There's little doubt that turning prospects into loyal customers and advocates is good for business. The question is: What does it take to achieve this goal? Unfortunately, it isn't easy. If it were, everybody would already be doing it.

Pushing customers up the path means delivering a positive experience each time. A positive experience is one that meets or exceeds the customers' expectations. On the rare occasions where their experiences don't go as planned, the organization must do whatever it takes to quickly make it right. Delivering positive customer experiences involves everybody in the organization. It's the reason the business exists. It's particularly important that companies focus on the people and systems that the customers interact with most often. The challenge for the organization is ensuring that all customer interactions are positive experiences for the customer. Why go through all that trouble, you might ask? Remember—there are not enough customers to go around.

COMPANY GROWTH AND CUSTOMER SATISFACTION

Although all businesses must make money to survive and prosper, making money should not be the sole goal of a business. The true goal of all companies should be to have the most desired products or services in the market at a fair price. Profits are important, but if a company pays attention to the customers' needs, the profits will follow automatically. The focus of any business should be customer satisfaction, and this focus should be a long-term focus.

Can a small nimble company that has always been quick to respond to customer needs grow very large and still maintain customer satisfaction? Yes, but it is harder, requires more vigilance, and depends on how the company handles growth. Keep in mind that customer satisfaction propelled the growth of the company.

How can customer satisfaction be maintained while achieving growth or after achieving growth? A simple answer is for the company to keep doing what it has been doing, but to do it on a larger scale. For example, say a company is in the grocery business, has one store, and has always given friendly, personal service to customers. The company's customer satisfaction has paid off, and it is now expanding from a small corner store to a large supermarket-type store. The company must keep the same level of services to preserve customer satisfaction. The company must keep the vegetables fresh, the store clean, and the aisles well lit, provide a service desk, train employees, especially the numerous newly hired employees, and emphasize the behavior they need to display. All of this will cost money and will cut into profit margins for several quarters. Short-term thinking might tell the business owner not to spend that much money on customer services, and, though the argument looks enticing, they need to watch out. Profits saved now may disappear as disappointed customers seek other stores. One disappointed customer can spread a tale of anger and disappointment to about ten other people and influence about five of them not to shop at your store. These potential, forewarned customers decide not to shop at the store, thus potential business is lost. Most businesses do not notice the slow draining away of their customer base until it is too late. It doesn't happen overnight. It is similar to a latex balloon blown up for a party. The balloon is fine during the party, but after the party, the balloon shrinks away to nothing day by day.

In another example, an entrepreneur in the restaurant business operates a small restaurant. Because of the restaurant's good food and cleanliness, people come from across town to dine at the restaurant. Soon, the restaurant becomes crowded and turns people away. The owner decides to open another restaurant and make the current assistant manager its manager. The manager was a good assistant manager when he was with the owner because he knew the owner was watching and wanted to please him. However, the manager isn't as industrious as thought. Also, it isn't his restaurant (investment); he just manages it. Because the owner can't be at two places at once, the food at the new restaurant develops a reputation for being mediocre, and the facility is not kept as clean. Because the restaurant has the same name, the inferior restaurant now is tarnishing the image of the original one. The point being made is that an expansion should be orderly and the training and organization of paramount importance. Businesses should expand at a pace that ensures the continuation of the established quality.

David against Goliath

How do small, individually owned (mom-and-pop) stores compete against the "big box" stores (Wal-Mart or Home Depot)? Can they compete? Should you open your own dress shop and compete against JC Penney, Palais Royal, or Sears? Should you open your own hardware store and compete against Home Depot, McCoy's, or Lowes? How can you compete against stores that have fifty times the floor space, a tremendous variety of merchandise, and large advertising budgets?

Billy Patton of Keyworth's Hardware in the small Gulf Coast town of Dickinson, Texas, has the answer (see **Figure 6.5**). Keyworth's has been in business over twenty years. Originally, his only competition included the franchises of McCoy's and Sutherlands Lumber. Recently, Home Depot, Wal-Mart, and Lowe's have built stores just a few miles away. Mr. Patton has tremendous competition from these giants, but, as he said, "Our prices

Figure 6.5   David against Goliath

roughly match our competitors, but we have superior service. As soon as a customer walks through my doors one of my floormen will approach the customer and say, 'Can I help you find anything?' Our customers never have to hunt for help. Plus, we'll special-order things for them and advise them on some repair problems."

Mr. Patton has it right. With so many stores selling the same things at nearly the same prices, all that is left to differentiate them is quick service and pleasing experiences in the store. The big box stores are impressive, but customers often have to hunt for help or wait while store employees finish with other customers. Buying a five-dollar item might take twenty minutes—and most people think that is fifteen minutes too long.

Training to Instill Company Culture

Earlier, the effect of training on expanding businesses was discussed. Training, along with the business culture, is extremely important. A company's culture should be customer-oriented. Trained employees are a shield against organizational deterioration. Training should also reinforce company culture and, where necessary, bring cultural improvement. People must have instilled in them a company culture that promotes teamwork and customer satisfaction. Both of these define the heart of the quality process. A good example is the United States Marine Corps, which uses training and tradition to create its world famous *espirit de corp*.

Expense of Customer Satisfaction

Quality cannot be achieved with hype, slogans, or money. Quality is not a program but a process. It must become part of the company culture. Money must be spent to keep the quality framework functioning properly and to ensure customer satisfaction. Does the expense of customer satisfaction cut into profit margins? Yes. Surveys, complaint departments, returned and restocked merchandise, and training employees to meet and respond to customer needs all cost money. Should organizations forget about customer satisfaction or at least cut back on it to maximize profits? Not unless they have an absolute monopoly and see no glimmer of competition for several decades. But the moment a competitor shows up, such companies had better be prepared to declare bankruptcy in a very short while. Customers do not forget or forgive easily.

Assume a large business has been doing very well and growing larger each year. The owner, secretary, and one other person have all been handling customer complaints, whether they are walk-ins or phoned in. But now the business is growing so fast that the owner and secretary just don't have the time to handle complaints. The company could hire someone, give him a desk and phone, and train him. That will be expensive, and that employee wouldn't even be producing a product or service for sale. That employee would not bring in any revenues—in fact, is consuming them. But the owner decides to do that because she knows one dissatisfied customer could cost the business a minimum of five potential new customers. If maintaining customer satisfaction costs a company 3 percent of net profits, is it worth it? Think about the alternative: no customers and no profits.

SUMMARY

Customer satisfaction is nothing more than satisfied customers. It's that simple, but not simple at all. Customer satisfaction is much easier to achieve for a small company than a large corporation. In a small business, management can more easily command and control the business and initiate needed training, and has more personal knowledge and relationships with more employees and a better knowledge of most if not all facets of the business. Large companies are severely challenged to do all of these things. Large companies are more impersonal and, as a rule, slow to change and to recognize the need to change.

Different surveying tools can measure customer satisfaction. Today, unlike any other time in history, the customer is king and has no sympathy for either the problems of the large corporation or small company. When customers pay for products or services, they expect nothing less than satisfaction.

The cost of a dissatisfied customer extends well beyond the business lost from that one customer. Typically, 95 percent of service time is spent addressing external customer problems; only 5 percent is spent on prevention. Gaining new customers costs five times as much as keeping current customers does. Customers must have twelve positive experiences to make up for one disappointing incident. The average customer tells nine to twelve other people about poor service or product quality, and that frightens away potential new customers.

REVIEW QUESTIONS

1. Define *customer satisfaction*.

2. List six ways to measure customer satisfaction.

3. List four variables used in customer satisfaction surveys.

4. Describe the two rules that guarantee business longevity.

5. Describe two things a company can do to keep customers satisfied.

6. Explain why customer satisfaction is more difficult to maintain as a company increases in size and expands its locations.

7. Explain why companies use training to maintain customer satisfaction.

8. List three costs of quality to maintain customer satisfaction.

9. Define customer *specifications* and customer *expectations*.

10. Why is having customers who are advocates more important than just having satisfied customers?

11. Describe how angry customers walking away affect future business.

GROUP ACTIVITIES

Students should divide into small groups and choose two of the topics below to discuss, and then report their conclusions to the class.

1. Role-play the owner of a sandwich shop at lunchtime. Your place is full of your normal customers; unfortunately, the delivery of fresh meat products by Deli Meats of Houston did not arrive at 9:30 a.m. as usual. Deli Meats has been your supplier for six years and has never missed a delivery. You call Deli Meats and find out their truck has broken down and they sent another truck to your shop, but it is stuck in a bad traffic jam caused by a five-car wreck. It is now noon and customers are leaving because they cannot get the sandwiches listed on your menu. Create a win-win situation with your customers and with Deli Meats.

2. Pick a well-known store that most of the people in your group have visited, and list the things you observed that the store does to keep its customers satisfied. Describe the importance of these customer satisfaction activities.

3. Make a list of customer *requirements* and customer *expectations* for a major department store (such as Sears or Target), and for purchasing a new car from a dealership.

4. Read the following article, "Revenge of the Irate Shopper," and then do the math using the following figures. A mass-merchandising store (a store similar to Wal-Mart) averages 180,000 customers a week. Each customer averages spending $130.00 a week. If 300 angry customers storm out of the store each week, (1) how much money did the store lose from angry customers resolved never to shop there again? and (2) how much money has the store lost from potential customers who have decided not to shop there based on the negative stories they have heard?

Revenge of the Irate Shopper

Written by Mike Speegle

Caveat emptor is a Latin saying that means "Let the buyer beware." The disgruntled shopper snarling at the store manager isn't the problem. The real and very serious problem is the customers who complain about the store to friends. A study shows that people told about a friend's or relative's bad shopping experience are up to five times as likely to avoid the store in question as the original unhappy customer. These tales of annoyance tend to be exaggerated with each telling. By the fifth rendition, the sales clerk who was just unresponsive has become abusive and insulting.

The survey of roughly 1,200 U.S. shoppers in the week before and after Christmas 2005, delivered some particularly bad news to the big box stores like Wal-Mart, Target, and Home Depot. It seems that customers of such mass merchandisers share their negative experiences with an average of six people. This is double to triple the audience sought by customers who've had negative experiences at other retailers. The biggest gripe about the big boxes is difficulty locating merchandise. Store managers at all types of outlets, meanwhile, are left out of the loop because irritated customers are five times more likely to vent to a friend than to a store representative. What retailers don't know (a complaint and how it is resolved) can really come back to bite into their profits.

CHAPTER 7

Employee Empowerment

Learning Objectives

After completing this chapter, you will be able to:

- *Write a definition of employee empowerment.*

- *Explain the difference between empowerment and development.*

- *Explain the difference between employee empowerment and involvement.*

- *Describe the importance of employee empowerment for a company.*

- *List four ways to facilitate employee empowerment.*

- *Describe Maslow's hierarchy of needs.*

INTRODUCTION

Quality starts with people. The human resource—the worker—is and shall remain the most important component of any TQM process. No matter how automated a process, a person is still required to operate, maintain, program, and monitor the process. The more complicated and expensive processes require a higher level of human talent. Once management finds the necessary talent, it must motivate them to become involved and accept empowerment.

Until very recently it was taken as a given that most workers were subordinates who did as they were told. Empowerment, the most important concept in TQM, has become necessary because employees must be empowered to make the needed process and organizational changes. The concept of empowerment is based upon the belief that employees need the organization as much as the organization needs them, and that leaders understand that employees are the most valuable asset in the firm. Participative management requires responsibility to and trust in its employees. Management must recognize the potential of employees to identify and derive corrective actions to quality problems.

Only through employee empowerment can the full strength of a company's workforce be used to sustain the company in its unending battle against competitors. All workers possess brains and talent. Management must stimulate and encourage their use.

EMPLOYEE EMPOWERMENT

Employee empowerment doesn't happen just because a manager calls his employees together and says, "From this minute on, you are all empowered. Go back to your workplaces and begin making improvements!" Empowerment cannot occur without employee motivation, training, and involvement. Employee empowerment allows employees to make autonomous decisions without consulting a manager. Thus, employee decisions can be small or large, depending upon the degree of power with which the company wishes to invest in employees.

Employee empowerment can begin with training (employee development) and converting a whole company to an empowerment model. Conversely, it may merely mean giving employees the ability to make some decisions on their own. When management offers employees choices and participation on a more responsible level, employees become more invested in their company and view themselves as representatives of the company. For employee empowerment to work successfully, management must be truly committed to allowing employees to make decisions. They may wish to define the scope or boundaries of those decisions. Building decision-making teams is often one of the models used in employee empowerment, because it allows for managers and workers to contribute ideas toward directing the company.

Making Empowerment Happen

Employee empowerment often also calls for restructuring the organization to reduce levels of hierarchy or to provide a more customer- and process-focused organization. Employee empowerment is often viewed as an inverted triangle of organizational power. In the traditional view of the triangle, management is at the top whereas employees are on the bottom; in an empowered environment, employees are at the top whereas management is in a support role at the bottom.

Frequently, the biggest changes required for empowerment to work must occur at the management and supervision levels. Turf protection, arbitrary rules, inflexible systems, capricious authority, poor listening, and reserving the right to make all decisions diminish the likelihood that employees will contribute even a fraction of their capability. True management skill involves the ability to direct, coach, delegate, and mentor individuals and teams, depending upon the situation and the employees' needs. Developing management and supervision with the skill and confidence to behave in this way is not a minor task. For this reason, many consultants recommend that employee development start at the top of the organization, with a consistent philosophy and approach backed up with observable behaviors.

Employee empowerment also involves management giving up some of the power they traditionally hold, which means managers also must take on new roles requiring more knowledge and responsibilities. It does not mean that management relinquishes all authority, totally delegates decision making, or allows operations to run without accountability. It requires a significant investment of time and effort to develop mutual trust, assess and add to individuals' capabilities, and develop clear agreements about roles, responsibilities, risk taking, and boundaries. In an empowered organization, people should not expect to be told what to do; rather, they should know what to do. The primary role of management is to support and stimulate their people, cooperate to overcome cross-functional barriers, and work to eliminate fear within their own teams.

EMPLOYEE DEVELOPMENT

Before most employees can be empowered, they must be "developed." Employee motivation and employee empowerment are part of employee development. Every work process eventually requires that people make decisions to do the right thing. For employees to act appropriately, they must be both motivated and trained. Motivation is a natural outcome of employee development and employee empowerment.

As humans, we are all created with a free will and the capability to make decisions. When employees do not make the correct decisions, problems soon develop, no matter how good the process or system. Every level of management needs to understand employee development and employee empowerment. There are an almost infinite number of small details that no one except the person actually doing the work can ever know. Without employee development and empowerment, an organization will have difficulty taking advantage of this valuable knowledge.

Usually an active employee development training program is required to nurture employee empowerment. **Development** consists of the training received when people are hired, and all future training as business plans and equipment changes. An employee with little training is an undeveloped employee. No matter how enthusiastic and motivated employees are, they do not possess the knowledge, skills, and tools to assist the company in continuous improvement. They are like hammers without handles—useless tools. In addition to training in basic problem-solving skills, employees should receive training that stimulates thinking and encourages them to make positive changes in their behavior, attitudes, and habits of thought about work.

Management should know which processes require training for employees to achieve goals, provide this training in a timely manner, and test the competence of employees to ensure effectiveness.

In autocratic organizations (see Table 5.2 in Chapter 5), employees have essentially no empowerment and no freedom to make even basic decisions. These same employees are community leaders and elected officials, serve on church boards, do volunteer work, run their own businesses, and in a variety of other ways demonstrate a capability far above what they use in their work. Imagine what could happen to businesses if they provided employee development and empowerment, and employees brought the same dedication, effort, and thought to work that they freely give away outside of work. Improvements of 25 percent to 50 percent in productivity have been demonstrated when employers develop and empower their workers.

Power to the People

Quality starts with people. One of the greatest underlying factors in the success or failure of any organization is the power of its people and how well that power is focused toward meeting the organization's objectives. As our society moves in the direction of automated processes and computer-controlled machinery, most of us tend to think that our reliance on people has decreased. However, all companies operate on the strengths and weaknesses of their employees. Employees have to design, maintain, and operate the systems that create output, which in turn generates profits. Organizations that can tap the strengths of their people are stronger and more competitive than those that cannot. Organizations that regard people as automatons or mere cogs in a wheel will never realize their full potential. In the long run, such company inefficiencies eventually destroy the company through an inability to compete.

Employees cannot become developed and contribute to empowerment unless they receive an active and ongoing training program. The pace of change today is daunting, especially in industries that rely on technology, such as the processing industry. Unfortunately, many companies view training as an expense that can be easily cut without a significant impact on company functions. As stated in earlier chapters, deciding not to increase the knowledge and skills of your most valuable resource is unwise, especially when you cannot operate a high-tech business with low-tech talent.

INVOLVEMENT AND EMPOWERMENT

A fundamental TQM precept is that employees must be involved and empowered. What do these terms mean? With **employee involvement**, every employee is regarded as a unique human being (not just a cog in a machine), and is involved in helping the organization meet its goals. Employees and management solicit and value each employee's input and recognize that each employee is necessarily involved in running the business for its survival. Sometimes employees do not want to become involved in the business of the company. They just want to show up and do their job, In cases like this management must force them to become involved through job enlargement, assigning them to committees and special teams, and so on.

Employee empowerment is a somewhat different concept. With employee empowerment, management recognizes that employees can identify and solve many problems or obstacles to achieving organizational goals, provides employees with the tools and authority required to continuously improve performance, and states its expectations about employees recognizing and solving problems, while empowering them to do so.

Facilitating Employee Empowerment

People talk about employee empowerment in many different ways, but the basic concept is to give employees the means for making important decisions correctly. When this process is done right, the results are heightened productivity and a better quality of work life. Employee empowerment means different things in different organizations, based on culture and work design. However, empowerment is based on the following concepts of job enlargement and job enrichment:

- **Job enlargement:** Changing the scope of the job to include a greater portion of the horizontal process. For example, with job enlargement, operators not only monitor and manage their areas of responsibility, but are also allowed to change out defective gauges and valves less than two inches in internal diameter and analyze unit samples.

86

- **Job enrichment:** Increasing the depth of the job to include responsibilities that have traditionally been carried out at higher levels of the organization. For example, with job enrichment, operators are included on quality committees, conduct ISO internal audits, and participate in process redesign and incident reports.

As these examples show, employee empowerment requires the following:

1. Training in the skills necessary to carry out the additional responsibilities
2. Access to information on which decisions can be made
3. Initiative and confidence on the part of the employee to take on greater responsibility

In an empowered organization, employees feel responsible beyond their own jobs, because they feel the responsibility to make the whole organization work better.

Most managers want their employees involved in improving the business or at least to be active participants in helping the business meet its objectives. However, in every organization, it is possible to identify people who make things happen, are well suited for the work they are doing, and enjoy their work, as well as others who seem to enjoy their work less, are perhaps not so well suited for the job either because of attitude or aptitude, or who are just along for the ride.

In chapter 15 of this book a quality tool called the 80/20 Rule will be discussed. This concept can be applied to a company's employees. We've all been in organizations in which 20 percent of the people in the organization do 80 percent of the work. Imagine what would happen in those organizations if everyone became as enthusiastic and productive as the 20 percent that do 80 percent of the work. That 20 percent would maintain their output, but now the other 80 percent of the people would increase their productivity. The mathematics indicates that an organization's output might increase by a factor of four or more.

Employee empowerment is a management philosophy that allows increases of this magnitude to emerge. Everyone works toward improvement. An executive at a major American electronics company said one person could no longer possibly run a major company by himself. The business world is too fast moving and complex. That executive wanted every one of his thousands of employees thinking about improvements. He considered a thousand heads to be better than his one. With a quality program and employee empowerment, the chief executive now has a lot more help. When employees are trained and educated, they need less supervision and guidance to do a better job, and they begin to solve problems rather than wait for someone to tell them what to do.

One of the prime jobs of supervision and management is to create the climate and the systems for employee motivation. Organizations need empowered employees involved from the neck up and not just from the neck down. However, even the best training and development programs cannot assure that all employees will get involved. Management must provide the opportunity and the means (see **Figure 7.1**). Then the employees must choose to take advantage of the employee development opportunity. Most employees, when they believe in and trust their management, will leap at the opportunity to make higher-level contributions to the organization.

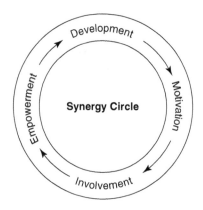

Figure 7.1 Synergy Circle

Round Pegs in Round Holes

Facilitating employee involvement requires recognizing the value of each individual, understanding human motivations, assigning people to positions in which they can succeed, and listening to employees. Consider an all-too-common scenario in which a marginally performing employee—the kind large companies only keep around until the next layoff—is reassigned to a new job and suddenly becomes a superstar. This employee is assigned to a value improvement team. Suddenly this employee's ideas on reducing costs through improved procurement strategies emerge daily. He starts visiting suppliers and shows them how to increase their productivity, which lowers their costs and prices while increasing their profits. He becomes a key member of a team that increases his employer's profitability, when only a few weeks earlier he was slated for a layoff.

This employee came alive because he was involved and empowered. He was placed in a position for which he had both aptitude and interest; his boss was interested in what he was doing and gave him the green light to fix problems. His boss didn't tell him what to do, he only told him what needed to be accomplished and empowered him to make it happen. The employee became an example of the productivity gains that result when people are assigned work they want to do and are given the freedom to do it.

This scenario has one caveat. In the processing industry, employees are hired for a particular position, such as an operator, which involves shift work. After one year, the employee cannot say to his supervisor that he could do a lot better job if he were working in the warehouse or as a firehouse technician—both jobs that do not involve shift work. The company spent thousands of dollars training this employee for the operating position and is not about to let him move to a new position just because he thinks he could do a better job or be happier there. There is an old saying: "Bloom where you are planted."

Suggestion Programs: Another Involvement Tool

New employees typically come to work enthused and wanting to contribute, but management must work diligently to prevent the erosion of employee enthusiasm and empowerment. One way of empowering employees is to ask them for suggestions. Many organizations use formal suggestion programs with varying degrees of success as a tool to facilitate employee empowerment. The format for these programs usually involves suggestion boxes and forms throughout the facility, with periodic management review of the suggestions and feedback to those making the suggestions. Many companies also have incentivized the process, offering cash or other awards for approved cost reduction or quality improvement suggestions.

Not every suggestion will be implemented, but all should be answered. This simple act of listening to employees raises their expectations for improvement. Listening and then failing to provide feedback on the status of a suggestion or improvement idea is probably worse than not listening at all. If managers don't listen, employees will only suspect that management doesn't value their ideas. If managers listen and then fail to provide any feedback, the suspicion will be confirmed.

Sometimes managers are afraid to provide negative responses to employee recommendations. Most employees are not offended by a rejection of their suggestions if the idea is not feasible, if the reasons for the rejection are explained, and is the explanation is offered in a constructive and appreciative manner. When suggestions are refused, a logical explanation for refusal is part of the learning process. However, when good suggestions are placed on the back burner for reasons that obviously do not make good sense, employees sense that management is not really interested in their suggestions. When that happens, management has just motivated the worker to do only his job and nothing else. Employees recognize that life is easier if they just keep their mouths shut and do their job. The significant contribution of employee empowerment, buy-in, and continuous improvement go out the window.

MOTIVATION FOR EMPOWERMENT

One of the most important tasks any management team faces is motivating its organization's members. Understanding what motivates people is an important element of empowerment. Without motivation, all the empowerment in the world will do no good, so it is necessary first to understand what motivates people. Understanding motivation is not simple because we all have different motivations.

Several models describe human motivation. The one that come closest to modeling human behaviors (in my opinion) is Maslow's hierarchy of needs (see **Figure 7.2**), developed in the 1940s. Maslow's model is based on an ascending order of human motivations, starting with the most basic motivation of survival, and then progressing upward through safety, love, esteem, and self-actualization. Maslow's model stipulates that, as each need is satisfied, people progress up through the hierarchy of needs. In other words, people faced with starvation or exposure would only be motivated by attaining those goals that would eliminate the threat of physical destruction. A person in this situation would not be too concerned about higher-order motivations, such as love or esteem. Once survival becomes a relatively sure thing, individuals move up the hierarchy of needs and take steps to assure their own safety. Once that need has been met, they continue to move up the hierarchy of needs to search for love, and then the esteem of one's associates, and then, finally, self-actualization (to be all that they can be).

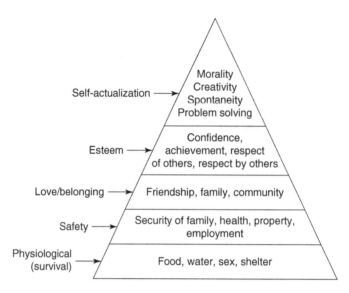

Figure 7.2   Maslow's Hierarchy of Needs

Often, individuals receive esteem and self-actualization from their jobs, through the facets of responsibility, challenges, and the camaraderie of teamwork. Many companies motivate through monetary rewards (bonuses) if certain goals above and beyond normal goals are met. A process unit may have four special goals, and a bonus of $1,000 per employee will be paid for each goal attained. Two other ways companies can motivate are through recognition and advancement.

ATTITUDE AND BEHAVIOR

Most people do not drive to work planning to do minimal work or determined to avoid any extra tasks sent their way. Those who do have this mind-set cannot be "truly" empowered because they have either become thoroughly disillusioned with the company or have a negative attitude resulting from childhood experiences.

The attitude that employees have at the workplace can be as important as their actual technical skill levels. Usually when an employee is said to have an attitude, it goes without saying that this refers to a poor attitude. Employees with poor attitudes are common concerns in the work environment. Actually, the attitude is not the problem; rather, the behaviors that result from that attitude are the real concerns. Employees with bad attitudes respond fairly typically, with attendance problems, marginal quantity or quality of work, interpersonal problems with coworkers or supervisors, poor communications, lack of cooperation in any activity, and so on. The list is remarkably similar no matter the job, company, industry, or part of the world. This connection between behavior and attitude is common. And, although employees cannot be terminated for their attitude, they can be terminated for the problems their attitude causes (behaviors).

SUMMARY

Empowered personnel have a sense of responsibility, ownership, satisfaction in accomplishments, power over what and how things are done, recognition for their ideas, and the knowledge that they are important to the organization. Without productive employees, the organization is nothing and can do nothing. Empowerment works best when employees need their organization as much as the organization needs them, and that need is much more than a paycheck and benefit package.

Experts also believe that TQM requires a company to review employee qualifications, promote technical skills development, and instill the spirit of teamwork. The human resource is and shall remain the most important component of any TQM process. Management must find ways to motivate its employees to become involved and to accept empowerment.

Employee empowerment is more than a management buzzword. It is a way of managing organizations toward a more complex and competitive future. A TQM strategy is deemed to fail if employees are not empowered. Quality starts with engaging the people responsible for processes and who know the processes the best.

REVIEW QUESTIONS

1. Why is employee empowerment important for a company?

2. _____ is the most important concept in TQM.

3. Explain how involvement is different from empowerment.

4. List three ways to motivate employees.

5. Write a definition of employee empowerment.

6. Explain the difference between empowerment and development.

7. Explain why development is important to employee empowerment.

8. Explain the difference between employee empowerment and involvement.

9. List three ways a company can motivate its employees.

10. List four ways to facilitate employee empowerment.

11. Describe Maslow's hierarchy of needs.

12. Explain how employee attitude affects a team and the work process.

GROUP ACTIVITIES

Divide into small groups and pick two topics to discuss and then report your discussion to the class.

1. Make a list of the experiences you had as an employee in your previous jobs when you were not treated as a customer, and explain how this affected your morale and attitude.

2. Using the same topic, explain how those experiences affected your or the company's productivity and efficiency.

3. Explain why employee involvement is critical to a company, and discuss some instances when you were involved or should have been involved.

4. Write an explanation of why you agree or disagree with the following statement: "Employees who are not treated correctly cannot be expected to treat external customers any differently."

CHAPTER 8

Teamwork and Teams

Learning Objectives

After completing this chapter, you should be able to:

- *Justify the purpose of teams.*

- *List four basic elements essential for a team's success.*

- *Describe the stages of team development.*

- *Describe the types of behaviors exhibited during each stage of team development.*

- *Identify four effective interpersonal skills.*

- *List four effective meeting management techniques.*

- *Describe the four effective meeting management techniques.*

- *Explain the importance of diversity to a team.*

- *Explain the importance of "win-win" thinking.*

INTRODUCTION

Teams have existed for hundreds of years, are the subject of countless books, and have been celebrated in many countries and cultures. Yet only in the last few decades has management become aware of the potential impact of single teams, as well as the collective impact of many teams, on corporations and companies. Teamwork is another important facet of total quality management, because teams, not individuals, do the work. A ***team*** is a group of people pooling their skills, talents, and knowledge to accomplish goals or tasks. According to a study of worker attitudes and behavior, change threatens Americans when it is imposed. Being involved in teams helps people accept this change.

When an employee becomes an operator, she is assigned to a crew (team) on a unit. She'll be one of four or five crew members responsible for running the unit properly, making on-specification product, and operating the unit efficiently and profitably. She cannot run a processing unit by herself; no one can. At times, she'll need help or be asked to help a fellow coworker. She is part of a team and will have to respond as a team member, not as a lone wolf individual. She may be on that team until she retires twenty-five or thirty years later.

THE BUSINESS NEED FOR TEAMS

Western society, especially American society, is very individualistic. However, many Americans enjoy being part of some type of team—sports, band, choir, and so on. If dedication to the team can be transferred to the workplace, individuals' objectives will be aligned with those of the organization, and the organization will be fully using the dedication and brainpower of all its people. With this base, the only additional ingredients necessary are the training and coaching the management provides.

Individuals using quality improvement practices can contribute to an organization, but rarely does a single person have enough knowledge or experience to understand everything that goes on in a process. Because the individual's ability to contribute is limited, major gains in quality and productivity most often result from teams. With proper training, teams can tackle complex problems and come up with effective, permanent solutions. Individuals can rarely do this.

Synergy of Teams

Besides pooling skills and understanding, team efforts have another distinct advantage over individual efforts—the mutual support that arises between team members. Quality improvement is hard work and takes a long time. One person's commitment and enthusiasm can diminish during a long improvement project, but the synergy that results when people work together on a project is enough to sustain enthusiasm and support through long, difficult times. When one member wants to give up, other team members act as cheerleaders who encourage that member to continue. As a spirit of teamwork invades an organization, employees everywhere will begin working together and become one large team.

Imagine the chaos that football, basketball, police SWAT teams, and the military would experience if one individual on the team decided he doesn't want to do things the team way, he wants to do things his way. These teams could not function, much less survive, without teamwork. The process industry also relies on teamwork. Each crew is a work team, plus special teams of individuals from various areas joined together to solve important problems.

DEVELOPING THE TEAM

Generally, most teams range in membership between two and twenty-five people. Team size varies depending upon the team's purpose, performance goals, complementary skills, and mutual accountability. The majority of teams have fewer than ten members. A larger number of people, say fifty or more, can become a team; however groups of such size usually break into subteams, because large numbers of people—by virtue of their size—have trouble interacting constructively as a group, much less agreeing on actionable specifics. Ten people are far more likely than fifty to successfully work through individual, functional, and hierarchical differences toward a common plan and hold themselves jointly accountable for the results.

Stages of Team Development

When a team is first put together, it moves through four stages of development (see **Figure 8.1**). These are the normal growth stages of a team, and rarely does any new team bypass these stages. The following are the four stages:

1. **Forming** occurs when a team is first put together. During this stage, everyone is polite and assesses one another. They're studying personalities and developing feelings for each team member, deciding if an individual is too wimpy, too aggressive, or thinks he is too good to be on the team. At this stage, members have a lot of suspicion, fear, and very little trust. Personal behavior is guarded, and many team members are tense in the new group. Members cautiously transition from individual to team member while determining acceptable group behavior.

2. **Storming** occurs when team members develop personality conflicts and attitudes toward others because the team is inexperienced, nervous, or frustrated with its assignment or progress. Individuals may refuse to objectively participate or may disagree frequently. Members may have arguments, express anger, and demand to be taken off the team. This is the most difficult stage, as team members feel the pressure of their assignments and are unsure of how to proceed. This is a critical time for a team, and a good conflict resolution manager or facilitator may be needed to keep the team from developing serious cohesive damage.

3. **Norming** is reached when team members become used to and accept one another, form ground rules, and know their roles. They establish a sense of group unity and positive relationships. They feel comfortable speaking out, making suggestions, or disagreeing with one another. Trust replaces suspicion and fear. In this stage, the team gels, and team members become more open, avoid conflicts, and have established group rules.

4. **Performing** occurs when all people know their jobs and are accepted as competent and trusted members of the team. They begin making progress on their objectives, and their confidence soars. They establish mutual cooperation and creative problem solving. They have worked out group and individual issues and developed team loyalty, pride, trust, and a willingness to go the extra mile. They are now an "A" team.

Common Purpose and Performance Goals

A common, meaningful purpose sets the tone and aspiration of teams. Groups of individuals become teams through disciplined action. They shape a common purpose and agree-upon performance goals, define a common working approach, develop high levels of complementary skills, and hold themselves mutually accountable for results. Individual team members do not fail; rather, when an individual fails, the whole team fails. Because team members are all in the same "boat," they look after and encourage each other. Management may give a team several performance goals that may involve ensuring safety, reducing operating costs, implementing continuous improvement, and so on. An employee on a process unit will have other goals, such as meeting production schedules and completing mandatory training. These are team goals that require teamwork. If one crew member does not complete her part, the whole team could suffer.

Figure 8.1 The Four Stages of Teams

Cookie-Cutter Teams

Cookie-cutter teams are made up of people who think and act the same. These teams have no diversity. Putting together a "cookie-cutter" team may not be wise, when it has been proven that individual differences add depth, create strength, and bring balance. A dozen drummers couldn't create much of a musical group, for instance. In the same manner, a team

96

of people with the same opinions, values, and viewpoints has less promise of crafting good solutions than does a more diverse group. Teams perform best when teammates bring a variety of abilities, experiences, personalities, and problem-solving approaches. For diversity to bring value to a team, management must take advantage of it and respect and use those individual differences to strengthen the team.

DIVERSITY

In the last thirty years, America has gone from a "melting pot" vision to a "salad bowl" vision. In the melting pot vision, all new immigrants were expected to drop their language, customs, and traditional clothes, and to blend in and become indistinguishable from those born in America. This did not make sense in a nation that values individuality and creativity. The salad bowl vision embraces everything that makes immigrants unique. We are all Americans in the salad bowl but we have kept our uniqueness, and this is symbolized in the "colors, textures, and flavors" of the salad.

People are different in their politics, religion, race, choice of music and food, interests, and so on. When a new person joins a team, he shouldn't be sidelined because he is different. Some people are reluctant to join a team because they think they look, think, or act different from the rest of the team members. All people must do their part to help the team identify and benefit from the diversity of its human resources (see **Figure 8.2**). However, members must be watchful for individuals who decide that being "different" means they deserve special treatment. Diversity can make teamwork seem more difficult at first, but it produces a more powerful unit. Team members must make conscious efforts to use the unique talents of everyone on the team.

BECOMING A TEAM MEMBER

One of the most basic requirements of effective teams is that the team includes capable team members. Teams need talent, not just bodies. The more talent a member brings to a group, the more that member can contribute to the team's success. By building their skills, team members also build the team's skills. Teams are powerful vehicles for personal learning and development. Their performance focus helps teams quickly identify skill gaps and the specific development needs of team members to fill them.

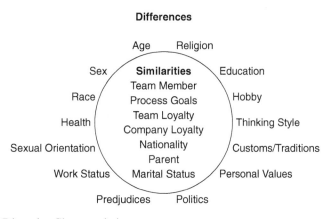

Figure 8.2   Diversity Characteristics

Developing Talent

A highly effective team is not possible without talented people. The weakest link in a chain determines the strength of the chain. The weakest member of a team may set the limit on what that team as a whole can achieve. Team members should constantly polish their skills and master the fundamentals of their jobs. They should strive for continuous improvement and encourage others to do the same so that the team never stops growing. Teamwork always takes a hit whenever people start to stagnate. Incompetent team members cannot be depended on. Members build high trust levels in the group when they bring talent and give teammates good reasons to believe in them. Members should understand what is expected of them in the way of duties, standards of performance, time frames and deadlines, and how to interface with others on the team.

Teamwork implies interdependence, meaning what members do affects others. On a process unit, all members of the work team are dependent on the others for the unit to stay on specification. A board operator cannot ignore the console, and outside operators cannot decide they will not do their duties one day. Operators on process units downstream of a process unit depend on the upstream unit sending them on-specification material for the success of their unit. What members fail to do can cause another team member to fail. Team members that are careless about covering their assignments may cause other teammates to abandon their duties in order to bail out the sloppy member. Sometimes a member will need to cover for teammates, because everyone needs a little help now and then. However, team members support their teammates best by covering their positions to perfection.

The best way to put a safety net under the team's performance is to be able to back up one another. Anybody can make a mistake, get overloaded, or just need a helping hand. Employees should know their teammates' roles or they won't be versatile as backup people. They should broaden their skills by cross-training so they have enough know-how to actually help. They should be willing to jump in and give a teammate a helping hand. By following these suggestions, they'll prevent a lot of team breakdowns, plus improve the odds that their teammates will be there when they need a backup.

Just having your name on the team roster doesn't mean you're earning your keep. Team members are supposed to make a *difference,* and making a difference means more than just showing up and doing only enough to get by. Talent alone guarantees nothing because how the talent is used is what matters.

"Me" versus "We"

Team members may experience the struggle of "me versus we" that arises when personal interests conflict with what's best for the team. Sooner or later, everyone comes up against a situation in which they must decide whether to look out for Number One or to make personal sacrifices for the greater good of the group. One person can never build a team by building up herself; she does it by helping her teammates perform at their highest level, by being a cheerleader, and pointing out their strengths and telling them when they do a good job. Make a habit of showcasing others. When a compliment comes your way, consider giving part of the credit to your teammates. By helping your teammates believe in themselves and building their confidence and self-esteem, one person can bring genuine growth to the team.

Personal Behavior

Trust is developed when people are tested. Only then do they get a chance to prove something. Will they keep their word? Do they honor their commitments? Are they consistent? Do they play fair? Can others count on them to help out when they are needed? Real teamwork requires people to have faith in one another, and the only way to build that faith is in the way they behave. Can you imagine working closely with a group of people you can't trust? In a climate of mistrust, the risk factors climb so high as to become a barrier to co-operative effort. Individuals start looking out for themselves at the team's expense because they doubt that the group will protect them. Everyone on the team should protect and nurture the trust level, because it is very important and very fragile. If building trust is a slow process, rebuilding it can be next to impossible. Once team members lose faith in someone, that person's reputation is shot.

Mavericks and Rebels

What do we do with those individuals categorized as mavericks or rebels? How do they fit into the team picture? They fit in when they make positive contributions to team goals and performance. To be a team member is not an argument for blind obedience or for mindless conformity. Being a good team player can mean having enough guts to take a stand against the leader when a point needs to be argued, an order questioned, or authority challenged when it is being abused. The key is to do it for the good of the group and its goals and not for personal gain. There is a difference between being a concerned team member versus a troublemaker.

TEAMS FOR CONTINUOUS IMPROVEMENT

As mentioned, natural work groups are teams, as are departments (such as operations, maintenance, administration), and quality circles. The following basic elements are essential for team success:

1. Team goals are critical and are the responsibility of all team members.
2. Good communication is high priority.
3. Mutual understanding, respect, and cooperation are emphasized.
4. One hundred percent team participation is mandatory.

The Team Approach to Problem Elimination

Doing business in today's world is just plain tough and grueling. Demands and problems never cease, and the complexity of some problems puts them beyond the power of any individual to solve adequately. Teams become a necessity for continuous improvement (see **Figure 8.3**). The skills and knowledge of team members make them the most likely sources of finding solutions to problems. Some business professionals balk at the use of teams because they don't like to "work by committee." However, using teams for problem elimination has the following advantages:

1. Teams can bring more talent, experience, knowledge, and skill to focus on problems.
2. Teamwork can be more satisfying and morale-boosting for people than working alone.

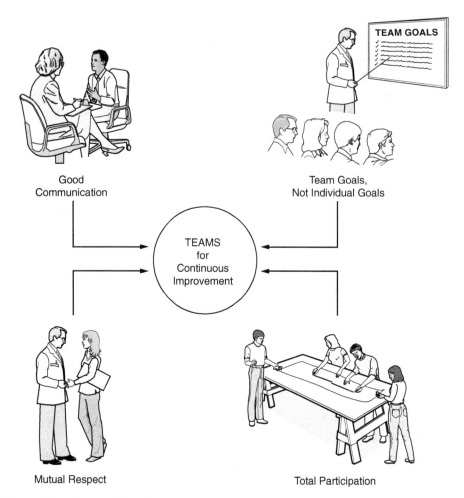

Figure 8.3   Teams Aid in Continuous Improvement

3. Team recommendations are more likely to be carried out than individual recommendations because people are more willing to support an effort that they have helped develop.
4. Teams can handle a variety of problems beyond the technical competence of any individual.
5. Teams can deal with problems that cross department and division lines.
6. When properly managed, teams can produce results quickly and economically.

For teams to be effective, two aspects of teams must be addressed: team maintenance issues and team support factors. Both are necessary for a successful team approach to continuous improvement and problem solving.

Team Meeting Management Issues

A team, like an automobile, requires maintenance to perform properly. Teams meet to decide issues and solve problems, and these meetings require energy and time. The team must address certain meeting management issues for each meeting to be productive. Some issues must be addressed constantly. The following issues are considered to be basic guidelines for good team meetings.

- Participant selection
- Agenda preparation
- Statement of objective
- Action assignments

Including these guidelines does not guarantee a successful team meeting, but ignoring them literally guarantees its failure. Although the team leader has the primary responsibility for these meeting management issues, all team members have some responsibility to attend to them.

Team meeting management issues are also important for formal teams, such as safety or quality teams to which crew members have been assigned. These teams are often also made up of members from other process units and departments. Such meetings should never be unorganized get-togethers. After only a few such meetings, team members may realize that team meetings are a waste of time, and they will lose interest in trying to be effective team members.

Participant Selection
A team may consist of more than just the crew members of a process unit. It might also have an engineer, someone from maintenance or the laboratory, and even someone with no knowledge of the process, such as an administrative clerk. The administrative clerk is valuable because of his "diversity," meaning he will think out of the box. Because his knowledge of the process is almost nil, he might come up with ideas that other members could never have conceived. Granted, many ideas may be useless, but one or two might be diamonds in the rough. Usually, only those people with knowledge and experience relevant to solving the unit problems are chosen when selecting individuals for team membership, but exceptions can be made. The team should be limited to between five and nine members, which will keep the team small enough to be manageable, but large enough for a good interplay of ideas.

Agenda Preparation
An agenda should be prepared and distributed to team members at least one day before the meeting. Team members need time to think about the agenda items and what data they can bring to assist the meeting. The agenda should include information about the meeting time and place, preparatory assignments (specific tasks to be completed by specific team members), and a roster of individuals involved. Supporting material should also be gathered and prepared for distribution.

Statement of Objective
At the start of the meeting, the team chairperson should state the meeting objective as explicitly as possible. This helps team members to understand the objective and focus their efforts on it instead of getting sidetracked. The objective should be restated whenever the team appears to be sidetracked.

Action Assignments
Team meetings don't solve problems, but they do produce strategies for solving problems. It is sometimes necessary to assign specific responsibilities (called action items) to team

members for implementing certain aspects of the strategy. They might be asked to bring certain data to the next meeting. It must be clear who has the responsibility to do what, and the time frame for implementation. This should be decided during the team meeting.

TEAM DYNAMICS

The interaction among team members is often referred to as **team dynamics**. If the team dynamics (see **Figure 8.4**) are managed properly, the team has a high probability that its problem-solving efforts will be successful. Success in team meetings is determined by the development of creative solutions to problems. Creativity is not an easy thing to tap. Although all people are creative to some degree, most of us have not learned to use our creativity because we have not been encouraged to do so.

All members of a team share some responsibility for encouraging creativity. Team leaders have very obvious roles to (1) encourage team members to speak their minds, (2) ensure that no one person dominates the proceeding, and (3) keep the meeting on schedule and on target. Team members have the roles to (1) share their ideas and opinions, (2) encourage other points of view, and (3) prepare for the meeting so they can contribute.

The following is a list of team dynamics issues that all team members should address:

1. Openness
2. Perceptual blocks resolution

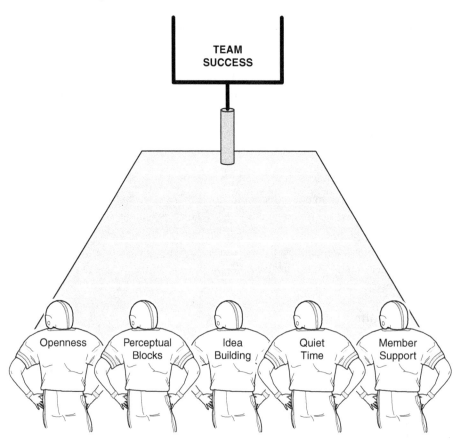

Figure 8.4   Team Dynamics

3. Idea building
4. Member support
5. Quiet time

The team dynamics issues affect the actual performance of the team. The better the team dynamics, the more effective the team.

With the **openness** issue, team members should encourage the free exchange of ideas. Ideas and proposed solutions should not be censored or criticized at the early stages of a team meeting. Team members can sort out ideas based on some type of criteria later, but should avoid doing so at the onset of the meeting because it discourages openness and spontaneity. To encourage openness and spontaneity, team members can record ideas as they generate them. These generated and recorded ideas may stimulate even more ideas.

A **perceptual block** is a preconceived notion about a situation or problem that prevents people from looking into potential solutions. With this issue, members may make up their minds and think they can do nothing about the problem. For example, for a long time, people who waited at credit unions or government agencies picked a line at a window and waited at that window until their turn for service came. People felt that they always got in a line that always moved slower than others, which seemed aggravating and inevitable. Then queuing theory was applied to the situation. Now credit union and government customers stand in one line and step up to the first available window that becomes free. Now there is only one line and it is always moving. The result is that customers are serviced much faster. Before this could happen, credit union and government officials had to change some perceptual blocks about how to run their businesses. Likewise, team members must try to overcome their perceptual blocks when they are considering ways to resolve a problem.

One approach for resolving perceptual blocks is to use a technique that is referred to as the *rolling why*? Whenever someone says, "We can't do that," ask "Why?" and follow up with more *whys*. The thrust of this technique is to encourage people to inspect the assumptions they make about a situation.

With **idea building**, everyone is expected to contribute in team meetings. Although not everyone has original ideas, just about anyone can build on ideas from other team members. By listing alternative ideas on a flip chart or writing board, team members have a chance to keep the ideas out front and in view. Idea building can take different forms, such as:

- Synthesis—A different application for an idea
- Innovation—The marriage of two ideas into a third
- Improvement—An idea remains unchanged, but the means of carrying out that idea may be improved

In addressing **member support**, no one person should dominate a team meeting, no matter how brilliant that person is. Domination does not foster participation or support. The greater the level of participation in a team meeting, the more likely all team members will support the final decision because they had a hand in developing it. As much as possible, it is important to create a win-win situation, in which no one loses or walks away angry or hurt. All walk away with their egos intact and feeling supportive of the team and its objectives. In a win-lose situation, someone loses and may lose loyalty to the team because of a bruised ego. The loser may tell other employees that the team is on the wrong track and that what they are proposing will never work. The loser may temporarily become nonsupportive of the team or divide the team's loyalties. If possible, always avoid a win-lose situation.

Quiet time is an important team dynamic, because many team members do not think clearly when under pressure. A good meeting policy is to allow some quiet time at the start of a meeting so people can gather their thoughts. The team leader can institute quiet time by asking members to be silent for ten minutes and to organize their papers and thoughts. This allows team members to prepare themselves for performing at peak mental capacity.

SUMMARY

Teamwork is crucial to the success of any quality program. However, not all teams are successful. Successful teams exhibit the following characteristics:

- They recognize that team goals are more critical than individual goals.
- They practice good communication.
- They show respect to all team members.
- They recognize that 100 percent team participation is mandatory.

Team development is accomplished in four principal stages: forming, storming, norming, and performing. *Forming* is the initial process, where team members meet and introduce themselves. The *storming* stage involves resolving conflicts and developing working relationships. By the time the *norming* stage is reached, team members have accepted one another, formed rules, and established roles. In the *performing* stage, all members know their jobs and are accepted as competent and trusted members of the team.

Good team members know their roles, are competent, and meet deadlines. They recognize diversity as a valuable asset. Each member also learns to be an effective backup for one or two other team members.

REVIEW QUESTIONS

1. Explain the purpose of teams.

2. Describe the advantages of a team over an equal number of individuals.

3. List four basic elements essential for a team's success.

4. List the stages of team development.

5. Describe the types of behaviors that would be exhibited during each stage of team development.

6. Explain what is meant by the sentence, "A highly effective team is not possible with low-talent people."

7. Explain how team members on any type of crew are interdependent.

8. Identify four effective interpersonal skills.

9. Describe two personality types that can be found on teams.

10. List four effective meeting management techniques.

11. Explain the four effective meeting management techniques.

12. Explain the importance of diversity.

13. Describe what win-win thinking means.

14. Explain the importance of win-win thinking.

GROUP ACTIVITIES

The object of this exercise is for students to practice using the information about teamwork and personal effectiveness in this chapter.

Divide into small teams and ensure is team is as diversified as possible, separating all "buddies." Each team should appoint a secretary and facilitator and try to avoid negative group dynamics during this activity by including the following:

- Ground rules
- Goals
- Agreement on a decision-making process

Have each team member role-play as a partner in a new start-up business. The team members can decide on what kind of business they have started up. Each has invested $25,000 in the business. Sample businesses are:

- Small grocery store
- Automotive service station
- Seafood restaurant

The team will decide on the name of the business and a logo. It will then decide on the tasks to be completed on a daily basis, who will do them, days off, rotation of tasks, pay, ordering supplies, and so on. At the end of the exercise, the team will present to the rest of the class:

- Their ground rules
- Their new business description
- An outline of individual tasks and who does them

CHAPTER 9

Communication

Learning Objectives

After completing this chapter, you should be able to:

- *List five different ways people communicate with each other.*

- *Explain how communication affects quality.*

- *List the three principal methods of communication.*

- *List the guidelines for written communications.*

- *Explain why* feedback *is important.*

- *Explain why listening is important in communication.*

- *Explain what* body language *means.*

- *Explain the importance of* gestures *in communication.*

INTRODUCTION

Getting a message across effectively is a vital part of individual success. As well, good communication is the lifeblood of organizations. Poor communication wastes time and money, creates bitter feelings, and may result in lost customers. Communication takes many forms, such as speaking, writing, and listening, and its purpose is always to convey a message. Communication also directly affects quality. Correct instructions, procedures, bills of lading, and exchange of information between workers are critical for customer satisfaction and for preventing waste.

The most frequent complaint at any place of business is the *lack of communication*. The frequency of this complaint implies that people don't talk to each other, don't e-mail each other, and don't write things down or pass on memos. Everyone *is* communicating, but they're doing a poor job of it. What they're saying and what is being *perceived* are not the same things; they're writing memos and e-mails, but the message *perceived* is not the message *meant*. As well, body language may be conveying a different message, or their rapport is so bad with their coworkers that nothing they say will ever be believed or accepted. Maintaining good communication is extremely important and extremely difficult.

COMMUNICATION

Good communication breathes the first spark of life into teamwork, and communication keeps teamwork alive. No other factor is as crucial to the coordination of efforts, nor plays as critical a role in building and preserving trust among teammates. Communication can be a make-or-break issue. The right hand must not only know what the left hand is doing, but must also know what the left intends to do because people need a sense of what's planned if they are to execute properly.

Good communication is essential to the safe and efficient operation of a processing unit. Technicians may be involved in scheduling turnarounds or other nonroutine duties, arranging equipment repairs with maintenance, discussing safety and environmental issues, quality issues, or process improvement issues. A technician who cannot communicate effectively will eventually find herself politely ignored by other technicians and management. Good communication, though seemingly simple, is not simple at all because too much is taken for granted. Good communication requires as much work and concentration as being a good ballplayer.

Communication requires three elements: a sender, a message, and a receiver. Surely something as simple as a person (sender) saying something (message) to someone (receiver) can't be that difficult. But many people who have been married to the same person for several decades have experienced how many times this simple formula hasn't been used correctly by one party or the other. Human nature interferes with these three elements and creates an amazing number of communication problems.

METHODS OF COMMUNICATION

People use the following three principle methods of communication on a daily basis: (1) written word, (2) spoken word, and (3) symbolic gestures and body language.

The written word is basic to literate societies, and includes letters, memos, reports, proposals, notes, agendas, and so on. It is the basis of organizational communication because it is relatively permanent and accessible.

The spoken word is effective only when the right people hear it, and includes interviews, meetings, conversations, debriefing, announcements, and so on.

Symbolic gestures and body language are positive or negative behaviors that are seen or heard. They include hand gestures, facial expressions, tone of voice, posture, and so on.

Of course visual images and multimedia are also forms of communication, but these are not normally used on a daily basis—except for the site intranet and Internet. Some are the domain of special departments in a firm or are contracted out (videos, television, posters, banners). Because of this, this chapter will only briefly address visual images and multimedia but not go into any depth.

Visual images include photographs, drawings, graphics, charts, logos, and so on. They include powerful conscious and unconscious messages. Multimedia is a combination of different methods, such as television, intranet, Internet, DVDs, films, and so on. Media are very useful when they allow recipients to respond and be imprinted by the message and/or to be interactive with the sender.

The Written Word

The written word, either via paper, e-mail, or online documents, is a primary form of communication because so many documents in the processing industry become, or may become, legal documents. A process technician will read and create many written documents. Written communication is required for many applications in the process industry, some of which are listed in **Table 9.1**.

Documents that are well written and easy to understand are composed by people who have clarified their thoughts, and more than likely, rewritten the document several times. Assume an employee and a coworker are told to write a procedure for a new piece of equipment. The document should not be a work of literature. Its purpose is to convey information, not to wow people with its vocabulary or writing style. The key to writing any document—whether a business letter, work procedure, or training method—is to write clearly and concisely. Use simple words and write straight to the point. Never use two words where one will do. Keep in mind the audience and write to the level of understanding of that audience. As an example, don't use engineering terminology and phraseology when writing for process technicians. The following are guidelines for written communication:

- Use short words (*said* rather than *announced*)
- Write short sentences

Table 9.1 Process Industry Written Communications

Memos and notes
Charts and graphs
Meeting minutes
Incident reports
Training manuals
Daily operating instructions and night orders
Work procedures
Permit systems
E-mail
Standard operating procedures

Table 9.2 Action Verbs for Procedure Writing

Activate	Count	Insert	Plug	Run	Throttle
Add	Decrease	Inspect	Prepare	Sample	Tighten
Adjust	De-energize	Install	Pressurize	Select	Transfer
Align	Depress	Isolate	Prevent	Send	Trip
Analyze	Discharge	Label	Pull	Send	Turn
Assign	Disconnect	Latch	Purge	Shift	Turn on
Attach	Drain	Limit	Raise	Shut down	Turn off
Block	Energize	Lock	Realign	Silence	Unlock
Bypass	Ensure	Log	Record	Start	Unplug
Calculate	Enter	Maintain	Reduce	Stop	Update
Calibrate	Evacuate	Monitor	Regulate	Store	Vent
Change	Fill	Notify	Release	Submit	Verify
Check	Flush	Observe	Remove	Supply	Wait
Clear	Guide	Obtain	Repair	Survey	
Clean	Identify	Open	Report	Switch	
Close	Increase	Perform	Resume	Tag	
Confirm	Indicate	Place	Review	Terminate	
Connect	Inject	Plot	Rotate	Test	

- Use active verbs (*open* versus *should be opened*) (**Table 9.2** lists common action verbs used for procedure writing.)
- Avoid jargon (closed valve versus deadheaded)
- Write the way you talk
- Revise when you have finished
- Cut unneeded words and sentences

Because others will read your document, it should be accurate and well formatted so that it is easy to follow. Depending on the type of document you are writing—(for instance, an operating procedure), you may want to use numbered paragraphs to make following it easier. That also keeps important points separate. Use headings for changes of subject and include subheadings if necessary. When you are done, have several colleagues proofread it for clarity and accuracy.

The Spoken Word

The spoken word can cause frequent misunderstandings because many people have very limited vocabularies and do not have the proper words to convey some ideas. A spoken message can also be misunderstood for the following reasons:

- The speaker is not clear about what he wants to say.
- The speaker is vague in how she says the message.

- The speaker's body language contradicts his message.
- The listener has decided in advance what the message is regardless of what the speaker is saying.

Poor oral communications not only lead to expensive or serious mistakes, but can also cause bitter feelings between coworkers, as each worker claims that was not what the other said.

Feedback   One way to prevent poor oral communication is to have the listener restate the message. Restatement by a listener is called **feedback**. Use the listener's feedback to avoid misapprehensions and errors. Feedback is essential to good communication, especially during critical times, such as a unit startup or an emergency situation. Feedback (1) verifies that the listener has correctly understood the speaker's message, and (2) allows the listener to react (show anger, enthusiasm, and so on) to what has been said, yielding information to the speaker.

HOW TO ASK QUESTIONS

Good questions are an important part of verbal communication. How a question is asked is important for establishing good communication. *Why, what, how,* and *when* are words that should be used often to seek answers. The right questions can open the doors to understanding. There is an art to knowing which question to ask and when. Keep in mind that the tone of voice in which a question is asked is a form of communication. A tone may convey anger, sympathy, or sarcasm, all of which can affect or influence the listener. It is best to speak in as natural or neutral a tone as possible to create a warm environment. The following are examples of two ways to ask the same question. Which one do you think will yield the most valid information and cooperation?

"Bob, what did you do wrong?"
"Bob, tell me what happened."

Listening: A Basic Involvement Tool

A man once commented that most people have two ears and only one mouth. The clear implication is that most of us might be well served by spending at least twice as much time listening as we do talking. Listening is one of the most effective tools managers use to promote employee involvement. This may sound trivial, but it is not. Listening to a fellow human is a powerful involvement tool. It helps the speaker feel that the person listening wants to understand what the speaker has to say. It encourages people to open up and to become involved. If no one listens, people won't become involved because they will recognize that no one values their opinions.

Listening is just as important as speaking because listening is 50 percent of any verbal communication. Most techniques for effective listening are really no more than common sense and good manners. The first, and perhaps most obvious, is to make eye contact while listening. When someone is speaking, listeners should refrain from speaking. Imagine what would happen if a manager asked an employee for an opinion on something, and, as soon as the employee started to speak, the manager jumped in and explained the real problem and what needed to be done to fix it. When this occurs, management is simultaneously telling the employee that his opinion is valueless, and that management has all the answers and does not really care what the employee thinks. Unfortunately, most of us can recall examples of such behavior. The great tragedy is that the people who understand the real problems and know what needs to be done to fix them often go unheard.

The next step in good listening is to ask questions, but to do so in a nonthreatening and open-ended manner. Open-ended questions are those that require more than a yes or no answer. Suppose a work center is producing an unacceptably high number of nonconforming parts. One approach to soliciting employee input in such situations is to ask employees what kinds of tools they need to make their job easier. This is a good approach, as it does not accuse the employees of producing nonconforming material. It clearly conveys management commitment to support the work center with whatever it needs to continuously improve, and it induces employees to speak up about needed improvements. Another approach—and one that should be avoided—is to ask, "What are you doing wrong? Why are you making so many nonconforming parts?" This sort of negative questioning is threatening, and few of us would be eager to share what we are doing "wrong" with management.

It's also a good idea to ask questions from time to time when employees explain their ideas for improving an operation to assure that what they are saying is understood. This tends to help speakers understand that the listener really is interested in what they have to say. One last thought on listening has to do with summarizing what was heard (feedback). This helps listeners confirm what the employee says and gives the employees an opportunity to correct any misperceptions or to provide more information. Summarizing in this manner at the end of a conversation further reinforces to the employee that management is committed to hearing what the employee has to say.

To facilitate listening skills, a listener should take statements at face value without searching for hidden meanings in what is being said. If a listener suspects the speaker's integrity, body language (such as evasive eye contact) and verbal signs (such as hesitation or contradiction) can provide clues to truthfulness. However, listeners should be careful to not hear only what they want to hear and to block out anything else. Some important points to remember about listening are:

- Listening intently inspires confidence in speakers, as they will sense listeners are interested in what they are saying.
- Misunderstandings are caused when listeners hear only what they want to hear. Listeners need to stay open-minded and focused on what the speaker is saying.
- Constant interruptions make it difficult for speakers to get across their points of view.
- Regarding what is said as trustworthy until proven otherwise will help prevent misunderstandings.

BODY LANGUAGE AND GESTURES

Body language is a common form of communication for many species. If you have dogs or cats, you are probably familiar with their body language: the cocked ear, the wagging tail, and the comical facial expression. Humans have body language with a range of unconscious physical movements that can either strengthen or damage communication. The movies give us classic examples: a bad guy putting his hand on his hips and sticking his chest out, or the heroine wrinkling the corner of her mouth and turning her head away. Even sitting completely still, a person can communicate her feelings by how she sits. As an example, a young man has angered his girlfriend, and she now sits on the couch, arms across her chest, lips pursed, back rigid. She isn't uttering a word but, boy, is she telling the young man something.

Body Language

Because of its subtlety, body language can be difficult to read and control, but it can be a reliable communicator of feelings. Most of us read body language unconsciously. Body language is a kind of covert language that adds to the spoken word, reinforcing and strengthening it. Body language is very hard to fake, which is why people become suspicious when someone's body language contradicts his spoken word. The way an executive enters a room sends an immediate message to the people gathered there. The way a good friend stands staring vacantly at nothing with shoulders bent conveys a message.

Postures convey feelings. Standing with hands on hips and a big smile while directly facing the person you are talking to conveys determination, interest, and confidence. A direct, unsmiling gaze and hands at the side indicate attention but neutral feelings. A body turned slightly away, evasive gaze, and a slight slumping of the shoulders indicates a negative attitude. Turning the body away while talking, facing sideways, and avoiding eye contact is a subtle snub, sending a message of reluctance to be involved.

Whenever possible when standing with people, leave a personal space of one yard between you and other people. Crowding too close creates an uncomfortable feeling and/or irritation. Introverts tend to keep farther away from people when talking to them; extroverts tend to move closer.

Gestures

Gestures, together with other nonverbal communication such as posture and facial expressions, are an important part of body language. Certain gestures go along with speech. Hands are used to shape sentences, to draw a picture in space, to strengthen or deny a statement. People increase the number of gestures they make when someone doesn't understand them or when they become angry or excited. People decrease their use of gestures when they are cautious or unsure about what they are saying. If a person is a liar or not a practiced liar, his gestures will decrease.

All skilled public speakers use gestures for emphasis. Making eye contact and nodding while someone else is speaking are gestures that show support or empathy. The following are other familiar gestures that connote the feeling of listeners:

- Raised eyebrows indicate interest.
- Making eye contact and leaning toward a speaker show readiness to assist the speaker.
- A hand around the throat and/or the around the body indicates the listener needs reassurance. He may feel threatened by the message.
- Closed eyes and pinching nose reveal inner confusion and conflict.
- Both hands open at chest level and spread sideways, palms up, sends a subtext of helplessness, a plea to be understood.
- One hand raised above the head emphasizes a point; two hands raised above the head signal triumph.
- Sitting back and listening to someone while steepling the fingers—holding the hands together at chest height with the fingers touching and pointing upward—

sends out a subtext of confidence. The higher the hands are held, the stronger the subtext of confidence, but they should not go above the level of the chin.

- Wagging a forefinger back and forth is a classic negative signal and sends a subtext of, "You are wrong."
- Pointing sends an aggressive subtext, and using an object to point with, such a pencil or cigarette, extends the threatening finger and makes the subtext more intrusive and aggressive.
- Standing with your hands on hips and thumbs back is another aggressive subtext that indicates confidence and toughness.

Gestures are often so unconscious that they can convey messages we are trying to hide. As an example, a CEO assured his employees that their jobs were safe and that the company was on a firm foundation, shortly before the company went bankrupt. As he talked, however, his left hand was clenched so tightly that his knuckles showed white. His employees heard his words, but they also understood the subtext of tension that his clenched fist conveyed. They went home, polished their resumes, and began mailing them out.

As a final note, there is a form of gestures used throughout the process industry and many construction industries. These are hand signals that guide heavy equipment operators (cranes, cherry pickers, and so on), and are universally used wherever excessive noise and hearing protection is required.

BARRIERS TO EFFECTIVE COMMUNICATION

Communication frustration exists in all businesses, especially the refining and petrochemical businesses. The opportunity for poor communication and miscommunication abounds. **Figure 9.1** reveals a partial list of process operator communication methods. Failure to clearly and precisely communicate on any one of these methods has the potential to create one or more of the following outcomes:

- Hazardous situation
- Loss of profit
- Off-specification product

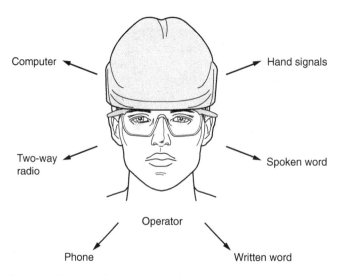

Figure 9.1 Operator Communication Methods

- Wasted resources
- Customer dissatisfaction
- Work stoppage
- Accusatory and emotional attitudes

People may communicate with each other on a daily basis, but what is said and the interpretation of what was said are often not the same. Learning to clearly and concisely state a message—either verbally or in writing—is a skill requiring the correct use of nouns, verbs, and the English language. A larger population reads Internet articles, but many of these articles are poorly written and shallow. Reading is an important tool for operators because they must read and comprehend many documents for necessary information regarding regulations and procedures. The following are other frequent problems in good communication:

- Incomplete information. For some reason, people omitted details.
- Excess information. People are swamped with too much useless data, and deciding what is important becomes difficult.
- Incorrect target group. The right information often reaches the wrong people. A lot of people are on distribution lists they shouldn't be on.
- Withholding information. Some people deliberately withhold information because they believe the information gives them more power. Withholding information can be a two-way process, both from top down or bottom up.

The form of the message (handwriting, use of an obscure phrase or word, and so on) may hinder communication. Also, the receiver may have barriers that block or filter the message. If receivers feel strongly or are biased about a topic, these feelings will affect how they interpret the message. When people communicate, they should share information and ideas with each other in a way that all parties can understand. Effective communication cannot flow from just one direction, which is why feedback and interaction are important.

SUMMARY

Communication is the lifeblood of any organization. In today's information-rich society, lack of communication results in many complaints and causes problems. Poor communication results from incomplete information, excessive and irrelevant information, assessing the incorrect target group, and deliberately withholding vital information.

Good communications require understanding, listening, and feedback. Communication requires three elements: a sender, a message, and a receiver. These three elements used properly result in good communication; when used improperly, they result in frustration, impeded results, and, often, hard feelings. Three common methods of communication are the written word, spoken word, and symbolic gestures (or body language). The key to effective communication is to keep it simple.

REVIEW QUESTIONS

1. List five different ways people communicate with each other.

2. Explain how communication affects quality.

115

3. List the three principal methods of communication.

4. List the guidelines for written communications.

5. Define *feedback*.

6. Explain why *feedback* is important.

7. _____ is 50 percent of oral communication.

8. Write a definition of *body language*.

9. Why is body language important?

10. List three barriers to good communication.

11. Explain the importance of *gestures* in communication.

GROUP ACTIVITIES

Divide into small groups and do both activities below, and then report the group conclusions to the class.

1. Make a list of real-life experiences with *gestures, body language,* and *filters,* and explain how they affected communication and you personally.

2. Draw several symbols or geometric figures connected by lines on a sheet of paper to make a symbol map. Give the symbol map to another group; however, do not show the map to one member of the other group. The other group, using oral communication only, will have that group member draw the identical symbol map on the board.

CHAPTER 10

Personal Effectiveness

Learning Objectives

After completing this chapter, you should be able to:

■ *Explain why personal effectiveness is important to the individual.*

■ *List five responsibilities employees have for their own personal effectiveness.*

■ *Describe why resisting change is futile for employees.*

■ *List four needs employees would need to excel at their jobs.*

■ *Explain how leaders promote an organization's mission and vision.*

■ *List four organizational aspects of a human relations system.*

INTRODUCTION

Employers all want "ideal" employees—those who are attentive, knowledgeable, great team players, and who go the extra mile. However, they get a certain percentage that is much less than ideal. Those who have been an employee of a large company will eventually encounter the term *high-maintenance individual*. This refers to a person who is a constant source of irritation and complaints, or who constantly complains to the human resources department about everyone and everything. These people are expensive to the company and disruptive to the workflow of their area. In cases where workers have been let go for being unruly, hostile, or simply awful to work with, the courts have held that employers need not tolerate misconduct on the job. Personality traits, such as an inability to work with others, are unprotected by law.

We live in a highly technical society, and that technology keeps changing. Those changes require people to constantly be involved in learning. Some people, once hired and trained for the job, decide they are through learning anything new. They are committed to themselves and their personal comfort only. These people reduce the efficiency of their company and their team. A significant number of workers like this are a threat to the efficiency and profitability of their company and process unit.

Lastly, some employees tend to blame the company for everything that goes wrong in their lives—even their home lives. They feel their lives are literally ruined because they have to work shift work, haven't been promoted to a supervisor, guys on the crew don't understand them, or they were given the wrong color hard hat. This chapter helps workers learn to make themselves more valuable employees and more responsible for their personal effectiveness and attitude.

EMPLOYEE RESPONSIBILITIES

"Well, I showed up for work on time. Now it's up to the company to get their eight hours of work out of me." I've heard this statement from several coworkers quite a few times. They were going to stand or sit there if their regular assigned work was done unless management walked up and assigned them more work. As one employee stated, "I'm no fool. I'm not going to do any more work than I have to." Their personal effectiveness (see **Figure 10.1**) was minimal, and their attitude was harmful to the company.

Commit to the Job

Almost every company in America—and many overseas—has reduced its workforce but not the workload. A crew of five employees has been reduced to a crew of four who now do the work of five people. Is it wrong for employers to expect more from employees? Is the employer trying to create a sweatshop? Or is the marketplace demanding more these days from all organizations? Competition is global and ferocious, as previous chapters have pointed out. Clients and customers constantly want better quality, better service, and cheaper prices. If a company can't meet its customers' demands, the customers will take their business to a competitor. Speed, too, has become essential because people have become

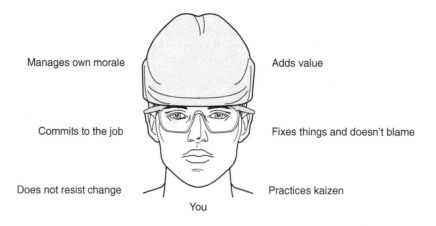

Figure 10.1 Personal Effectiveness Characteristics

used to having everything instantly. The only way most organizations can even hope to compete is to employ high-performance employees.

Before the world became so competitive, the common solution to problems was to hire more employees and spend more money. Today, that approach is equivalent to driving nails in the company coffin. Instead of throwing more people at problems, organizations today now employ fewer people. Companies must do more, faster and better, with less. This calls for highly committed and skilled people. Companies cannot afford to retain employees who merely put in their time and give only half-hearted efforts to do their jobs. All teams involved in competition—football teams, combat teams, production teams—cannot afford half-hearted players and cannot win without total commitment from all team members.

Don't Resist Change

Everything today is changing fast, and whatever task an employee is doing today could be different in six months. Companies must change if they are to survive. Technological evolution requires companies to change or go out of business. People must also embrace change. Gone are the days when years pass with no discernible difference in people, processes, or output.

In today's highly competitive environment, companies are constantly being bought, sold, or merged. When I was completing this book, two chemical companies and one refinery were changing ownership. I experienced change when British Petroleum bought out the Amoco Corporation. Career success belongs to the committed that invest themselves sincerely in their jobs. Recommit quickly when change reshapes your work. If you find you can't recommit to the culture of the new company, polish your resume and move on. Don't waste your energy resisting change because change is inevitable. Either buy in or move out because that is best for both you and your new employer. Everyone is expendable, even the company CEO.

This is not the same as asking that you be loyal to the organization no matter how unfair the organization is. The world today puts hard limits on how loyal employers can be to employees in return. All companies have to meet a bottom line, and sometimes they meet it by selling process units or laying off people. Investors and their boards of directors will demand that of them. It is in an employee's best interests to adapt, support the new company, and show job commitment. Strong job commitment makes work far more satisfying. It also empowers people, bringing out their very best potential, and making them more valuable employees.

When a company is under new management, the new company culture may offer more freedom than employees prefer. For example, if an employee has found comfort in having other people supervise him and stand accountable for problems and results, he may start to sweat when this responsibility becomes his. On the other hand, when an employee behaves like she's in business for herself, this gives her the chance to be noticed as an exceptional individual and will give the mind-set that will serve her best in years to come. Organizations simply can't guarantee peoples' careers like they did in the past when a person worked for one company for thirty years. It pays for employees to behave like they are in business for themselves because they are making themselves flexible and nimble, just in case things change for the worse.

Changing a team demands the most difficult learning imaginable—which is unlearning. The new company may do things differently. The company does not care that employees have done things a certain way for fifteen years. It demands that they give up those habits and ways of doing things and now do them a new way. Don't resist. Because the new company had the money to buy out the old company, maybe they know something the old company didn't know.

Add Value

Employees need to add value to their companies by contributing more than they cost. Employees often mislead themselves and assume they are entitled to keep their jobs if they are merely responsible and do good work. Some employees even have the idea that they have the right to coast along until retirement after being employed for twenty years or more. Thirty years ago, these assumptions might have been true, but not today. People who have shown true devotion to a company over years should be recognized, but must also realize that they still need to add value. They should not confuse longevity on the job with devotion to the job. The fact that a person has been on the payroll for years means nothing. Their contribution counts, not the years they put in. Many people stay busy and work hard without adding any real value to a company. Their careers are built on false assumptions. They will be better off if they think in terms of being paid for the value they add rather than for their tenure, good intentions, or activity level.

Practice Kaizen

A financially successful organization can protect the longevity of an employee's career. If a company improves in the way it does business and increases market share, an employee's future is usually more secure and has a better potential for advancement. However, keep in mind that the organization cannot improve unless its people improve. Continuous improvement—what the Japanese call *kaizen*—offers some of the best insurance for both employees' careers and the organization's success. Kaizen (pronounced *ky-zen*) is the relentless quest for ways to do things better, less expensively, and to make higher-quality products. Think of kaizen as the daily pursuit of perfection. Kaizen keeps employees stretching to outdo the previous day's performance. Continuous improvements may come bit by bit, but enough of these small, incremental gains eventually add up to valuable competitive advantages.

Without kaizen, employees and employers will gradually lose ground. Eventually, they'll both be out of business because the competition never stands still. Good quality by itself is unwise because a company's survival depends on its improving quality, getting better at a faster rate than its competitors. If a company does not get better faster than its competitors, then that company will lose the competitive race. It competes not just against other American companies, but also against companies all over the world, and they want that company's customers. "Good" quality today is mediocre quality tomorrow.

All employees should assume personal responsibility for upgrading their job performance. Their productivity, response time, quality, cost control, and customer service should all show steady gains. And their knowledge and skills should constantly increase. Granted, this drive toward an ever-improving performance doesn't guarantee job security, raises, or promotions. Employees can still be victims of circumstances. But if employees passionately practice kaizen, they will build a high competency level that will make them very desirable to the job market.

As an employee, you might question why you should make so much effort toward kaizen and increased competencies, knowledge, and skills when some of your coworkers do not make the effort. That is their mistake—don't make it yours. Just as you practice kaizen on the job to make your work process more successful, practice kaizen to make yourself more valuable. Protect your future by developing your talent. You don't want to be a fifteen-year employee that management performance reviews evaluate as being "okay." That could be interpreted as you meeting the minimum requirements for the job, which can be interpreted as not adding value.

Fix Things, Don't Blame

Problems are the natural offspring of change, thus we will all see plenty of them in the years to come. It is important for every worker to be recognized as a problem solver. Organizations need people who solve problems, not point them out. Too many employees get this confused and think complaining is a constructive act. They quickly identify the problems but contribute little toward improving things. Their attitude is that management is responsible for making and keeping things working. That was true twenty-five years ago when there were several levels of management, but not today when organizations have decreased those layers.

If you are an operator on a process unit in a chemical plant, you will have a chemical engineer who is the manager of the unit. The engineer will show up about 8:00 A.M., look at the logbooks, ask questions about how things went during the night, then leave to tackle a stack of papers on his desk and attend several meetings. Before his day is over, he will make one more visit to the unit, issue a few night orders, and then go home. You, the operator, are running the unit, not the engineer. You know best the problems the unit has and have the suggestions for eliminating those problems. If you wait for the engineer to solve your problems, they will never go away. An organization's value grows out of its individual employees' values. The organization's results are merely an accumulation of its workers' results. The worker, instead of being a finger-pointer, should assume ownership of problems and let the solutions start with him.

Manage Your Morale

Take responsibility for your attitude. Organizations want employees who can cope with change without breaking stride. Rapid organizational change guarantees that almost everyone is going to meet with some disappointments throughout their careers. When employees' attitudes sour because of company changes, they often blame the company, even if economic conditions forced the change. Nobody is well served by this line of reasoning. It doesn't matter that some organizations and managers treat their people unfairly at times. Many employees don't deserve high marks for how they treat their company equipment and resources either. Higher management can end up doing things that are hard for people to accept, but they may be things necessary for the company to survive. Keep in mind that the world at large displays no concept of fairness in the way it deals with organizations.

Employees can be bitter about how their careers are affected, or they can demonstrate their resilience and ability to handle disappointment. Management is not responsible for maintaining their positive attitudes. Anyone waiting around for management to heal a wounded spirit will be hurting a long, long time. Employees should do what is best for their careers, depersonalize the situation, and accept the changes as the luck of the draw. They should avoid harboring resentment toward managers. They are merely doing what their manager has told them to do, and their manager was only responding to the dynamics of globalization and competitive pressures.

Hold Yourself Accountable for Outcomes

Organizations are insisting on new levels of accountability in their employees. Responsibility, power, and authority are being pushed down to the lowest levels. Twenty-five years ago, an operator in a petrochemical plant or refinery only operated equipment—period. Today, they may do preventive maintenance, predictive maintenance, serve on quality and safety committees, order supplies, and train new employees. Careers carry more personal exposure these days. A company—and individual—survives on acceptable results, not excuses.

In these times of self-directed teams, empowered employees, and organizations without boundaries, your worth as an individual employee will also get measured by the work group's collective results. Holding yourself personally accountable for outcomes requires that you consider the big picture. Look beyond your own immediate behavior and the specifics of your job description to see if you're really doing all you should to bring about the right results. Those in operations, maintenance, technical support, and shipping departments, and so on, are all in the same boat. They all need to learn to work across departmental boundaries and to combine their efforts seamlessly with others who are contributing to the same end results.

MOTIVATION

Enthusiasm makes an ordinary person extraordinary. As individuals, we all have the same needs that must be fulfilled if we're going to excel at our jobs. These needs include:

- Economic security—We must feel that we're getting a fair day's pay for a fair day's work.
- Personal esteem—We all want to be viewed as value-added elements to the organization. No one wants to be thought of as just average or below average.
- Personal worth—We need to feel that we're contributing to a worthwhile goal. Learn the goals of your organization.
- Personal contribution—We want to be listened to and have our ideas heard.
- Personal recognition—We all need feedback. Managers need to communicate about the good work (as well as the bad) that their employees do.
- Emotional security—We all need to be able to trust the managers we work for and feel that they'll be honest with us. Employees are adults and want to be treated as such. Expect to be kept informed about the highs and lows of the business.

Employees are not naive and do not expect the world to be fair, because they know it is not. No matter where you are in the world, there will always be some people you consider incompetent and who don't do their fair share. Don't let this discourage you. Wise employees accept and understand this, making the best use of their talents and opportunities to forge positive attitudes, personal dedication, and commitment to success.

LEADERSHIP

All businesses need good leaders. A good definition of leadership is "the observable behavior that makes people want to follow or emulate another person." A leader is a person who can lead by example, persuade others, communicate both upward and downward in the company hierarchy, empathize with customers and subordinates, and motivate people with confidence. There is a shortage of good leaders. Can a worker such as an operator in a refinery become a leader? Yes, there is always upward mobility for people with ability.

Machine operators or workers on an assembly line who diligently and enthusiastically work to eliminate waste can serve as role models and leaders for their peers. Management recognizes and supports such leaders of continuous improvement, whatever their positions in the organization. This support encourages others to follow their example. An organization should have a formal and informal reward system so that people perceive that becoming involved in continuous improvement is in their interest. They should perceive that personal continuous improvement and process continuous improvement is the only acceptable way to act in the organization.

The most important part of any business or organization is the interactions of people. If you desire to become a leader or are already a leader, show everyone at every level in the organization courtesy and respect. Everyone (including janitors, delivery people, and so on) deserves to take pride in what they are doing. Good leaders truly respect and value others; in return, others will try to help them. Education should not be equated with intelligence. Mechanics and janitors who may not have had the opportunity or the money to go to college might make fine leaders. A college degree may give an employee an edge. However, as Dr. Deming said, a degree is a "learner's permit" at best.

Leaders promote the organization's mission and vision. They work to get everyone's support for the vision, help determine what needs to be done to achieve it, and energize team members to do their parts enthusiastically in making the vision a reality. Leaders must be enablers for teams. They help provide the necessary resources, including education and training, and help remove any barriers. Leaders empower others, giving them the authority and the power to effect change in the process. Leaders in the new system not only encourage people to take initiative, but they also actively promote change. Change means uncertainty, and people must have a vision of the better future for them to bring change. Leaders communicate the vision of the organization at every opportunity.

ORGANIZATIONAL ASPECTS OF THE HUMAN RELATIONS SYSTEM
The success of a human relations system for personal effectiveness and continuous improvement depends as much on an organization's policies and procedures as on managerial behavior. People cannot work in a new system without the full support of the organization. Through policies and procedures, an organization determines such things as pay scales, supervisory levels, and promotions. These policies and procedures play an important role in determining people's attitudes toward their work and the organization. The following are essential organizational aspects of a human relations system:

- Rewards
- Advancement
- Recognition
- Education

Each element should reinforce the goals of the management system and the changes it seeks to implement. Managerial actions speak louder than words. It is not enough to say that people should work in the new way. Written and unwritten policies and procedures should demonstrate that working in the new way is best for the individual as well as the organization. Through its human relations system (explained in more detail below) both formal and informal, management shows its sincerity about wanting people to think, talk, work, and act in a certain way. The essential organizational aspects of a human relations system are described in the following paragraphs.

Rewards—Both monetary and nonmonetary rewards retain their importance in the new system. Monetary rewards include pay, bonuses, stock gifts, and merit raises. Nonmonetary rewards include special prizes, dinners, plaques, tickets to events, and all the other ways organizations show their appreciation for individuals or teams.

Advancement—People learn what the organization wants by observing the behaviors of people who are promoted. Therefore, management that is serious about working in the new way promotes workers who have joined the team in the battle against waste. As well, workers who are most committed to change and continuous improvement climb the management ladder quicker than those who are not involved. If people see that striving for continuous improvement is a major criterion for advancement, they will quickly realize that working in the new way is in their self-interest. And if people understand that teamwork is expected as part of the job performance, they will more readily share credit for accomplishments.

Recognition—People want their efforts and contributions recognized. That desire is a powerful motivator. Too often, management recognizes accomplishments with money when people crave a word of praise. Ways to give nonmonetary recognition include:

1. Saying *thank you*
2. Arranging a presentation for top management of a successful project by the participants
3. Arranging for write-ups in the company newsletter
4. Arranging for awards of some sort

Education—Education (the "know-why") and training (the "know-how") are important human relations tools for overcoming barriers that prevent change and continuous improvement from being implemented successfully. People must be taught why working in a continuous improvement system is central to their self-interests and the good of the organization. For education and training to be effective, they must be integrated into all phases of the business. For example, many companies offer courses on technical tools, statistical control, and participative management. Unless and until those courses are tied directly to working toward continuous improvement, they lose a great deal of their value. People have to understand why they are going to training and what the training is expected to do for them. They need to understand that they are participating so they can help work successfully on projects to find and get rid of waste, for example.

SUMMARY

Personal effectiveness is important to the individual, and the individual is responsible for achieving that personal effectiveness. Employees are responsible for developing their knowledge and skills and for their attitude and morale. One of the hardest things individuals will be asked to do in their lives is change. If they have done things a certain way at a company for which they have worked twenty years, and they are asked to do things differently when the company is bought out, it is futile to resist.

Employees should not resist change, should continuously improve their talents, and should fix things instead of complaining about things. They should hold themselves accountable for their outcomes and their motivation. It is not the company's job to keep their morale high.

The company must supply the human relations needs and conditions that motivate employees to excel. By doing so, a company can truly achieve the buy-in and full empowerment of their workers.

REVIEW QUESTIONS

1. Explain why personal effectiveness is important to the individual.

2. Describe how you could increase your value as a process operator.

3. List five responsibilities employees have for their own personal effectiveness.

4. Describe why resisting change is futile for employees.

5. _____ means continuous improvement.

6. Give three reasons why employees should develop their talents.

7. List four needs employees require to excel at their jobs.

8. Explain how leaders promote an organization's mission and vision.

9. List four organizational aspects of a human relations system.

GROUP ACTIVITY

1. Read the following article, and be prepared to discuss it.

Play Nice with Others and We'll Hire You

Written by Mike Speegle, March 2008

You may have the right credentials for the job, but the wrong personality. A résumé and a brief job interview can't answer the question that matters most to a new employee's coworkers: Is this person going to make a good teammate or be an absolute pain? Despite a serious labor shortage in many sectors, some employers are pickier than ever about whom they hire. Many businesses are stepping up efforts to weed out people who might have the right credentials but the wrong personality. This is called the "plays well with others" factor. A few years back Robert Fulghum wrote an extremely popular book titled, *All I Really Need to Know I Learned in Kindergarten*. In the book's central theme, humans learned very important rules in kindergarten, such as *be nice to each other*, *share things*, and *help each other*. Employers are seeking employees with those personal characteristics. When you were young and playing in the schoolyard or playground, remember how someone was always a bully, didn't want to play by the rules, or whined about everything? Well, those people have grown up and are entering the workforce, and management doesn't want the headaches and problems they cause.

In the past, management conducted the interviewing and hiring of new employees, but many companies today assign three or four employees whom the new worker will be working with to the interview team. The interview process has become more time-consuming and sophisticated. Job candidates at investment banks have long endured dozens of interviews designed, in part, to see if new employees will get along with everyone they'll work with. Many of the interview questions are sophisticated and have psychological nuances that reveal important things about the psychology of the interviewee. Many companies are setting up higher hiring hurdles. With the national unemployment rate low, at 4.7 percent, and the baby boomer generation heading into retirement, employers from GE, British Petroleum, and Microsoft to even rural hospitals are worrying about finding enough workers with the appropriate knowledge and skills. Part of the skill set they are seeking is good interpersonal skills—playing well with other people.

Some companies have turned the hiring process around. Instead of looking for reasons to hire a person, they are looking for reasons *not* to hire a person. They are doing this because they would rather reject one or two good people who would have made great team members than hire one bad person. Why all the fuss about hiring someone who is a constant irritant? Can't they just be fired? Yes, they can, but this may become very time-consuming and expensive as management documents incidents that are grounds for termination. Plus, the terminated individual may file a lawsuit against the company, which the company will probably win, but that will cost the company money.

Some companies have developed work teams that have bonded so well that they meet and help each other on weekends or after work. For example, the team members of one company all showed up to help a team member paint her house. This is how closely knit and committed to one another the team members are. In a search to find the right person to join this type of team, several companies have an eight-hour job interview process. The intent is to wear away any fake pleasantness of an interviewee to reveal the real person. After five hours of being interviewed, the interviewee may no longer have the stamina to keep smiling and deceiving the interviewers.

A senior human resource manager at one cruise line said, "You can teach people any technical skill, but you can't teach them how to be a kindhearted, generous-minded person with an open spirit. That is the personality you have to have to work onboard a cruise ship." This cruise line did not brag about its wonderful benefits packages to potential new employees. Instead, it sends job applicants a DVD showing two filmstrips of a crewmember cleaning toilets and of a dishwasher talking about washing 5,000 dishes in one day. Be prepared to work your butt off, another DVD warns. The DVDs are meant to scare off potential employees who cannot accept this type of work, and it works. After watching the DVD and hearing a truthful description of life onboard this company's cruise ships, the majority of applicants turn down the job. New employees then undergo a drug test and a physical examination. They have to pack up their lives for so many months, are bought plane tickets to the cruise docks, and are outfitted with hundreds of dollars in uniforms. The company loses a lot of money if their new employees get on board and say, "Hey, this is not what I expected! I quit!"

In 2005, I attended a process technology conference in Houston. A senior human relations manager for a major integrated oil company told the college representatives present that the oil company could teach new employees how to operate equipment and to understand the dynamics of the process, but they couldn't teach them about interpersonal skills. The manager asked the college representatives to put more emphasis on teamwork and interpersonal skills. Too many new employees were being terminated during their probationary period because of poor interpersonal skills. The human resources manager said maybe too many young people watched too much of *Beavis and Butthead* and the *Simpsons*. That kind of attitude wasn't going to cut it in the processing industry.

At a nonprofit company that builds playgrounds, the board of directors confronted the company's CEO over the organization's high employee turnover. The CEO rationalized that his employees were on the road too much, when in reality, they were the wrong people in the wrong roles. The CEO began to think about whom left, and why, and then focused on the characteristics of workers who stayed. Those who stayed were very quick and very smart, fit into the team, and had can-do and will-do attitudes. His team began to keep a closer eye on job applicants in the reception area, which is set up as a playground, to see how they acted around playground equipment. The CEO said, "The people who came early may have to sit on a swing or the bottom of a slide. People who remain standing with a tight grip on their briefcases instead of sitting on the playground equipment aren't asked back. It is clear this isn't the job for their type personality."

CHAPTER 11

The Economics of Quality

Learning Objectives

After completing this chapter, you should be able to:

- List four economic issues that all national leaders face.

- List five goals of economic systems.

- Discuss the relationship of supply, demand, and price.

- Discuss the two factors that work together to determine the supply of a product or service.

- Explain how competition drives an economic system.

- Explain why governments regulate competition.

- List six factors that influence productivity.

- Discuss how information technology has affected the global marketplace.

- Explain why the customer has more "power" today, and how that power influences companies.

INTRODUCTION

Most of the ideologies of the modern world have been shaped by the great economists of the past—Adam Smith, David Ricardo, John Stuart Mill, Karl Marx, and John Maynard Keynes. It is prudent practice for world leaders to receive and launch the advice and policy prescriptions of economists. Economic theory has created a species called *Homo economicus,* a supposedly rational creature who grabs as much as he can for himself. In other words, ordinary people are keeping an eye on their wallets.

The president of the United States benefits from the ongoing counsel of his Council of Economic Advisers. Why are political leaders concerned about economics, especially in democracies? Because as long jobs, decent wages, and advancement opportunities are available, the citizens of the country will tend to be satisfied with their status quo. If jobs aren't available and wages are low or decreasing, citizens seek change, either through the ballot box or revolution.

The broad spectrum of economic issues that political leaders face includes topics such as unemployment, inflation, economic growth, productivity, taxation, public expenditures, poverty, the balance of payments, the international monetary system, competition, and antitrust.

WHAT IS ECONOMICS?

Economics is the production, distribution, and consumption of goods and services. The economy of a nation, whether it is growing, shrinking, or just standing still, is always on its citizens' minds because it affects their **standard of living**—the minimum necessities or comforts held to be essential to maintaining a person or group in a customary or desired status. People are concerned about maintaining or improving their standards of living. This is hard to do in a sluggish or shrinking economy. The standard of living is affected by a nation's productivity, and productivity is determined by a worker's output per unit of time. If Joe Doe made four widgets per hour the previous year, and he is making five widgets per hour at the same cost this year, then his productivity has gone up. Productivity is dependent on skills, talents, morale, and initiative.

An important branch of the social sciences, economics focuses on three critical areas of human behavior. First, economists study the problems that arise because resources are scarce and human wants are unlimited; therefore choices have to be made. Making these choices forms the other two areas of human behavior that economists study: allocating scarce resources to produce goods and services to satisfy insatiable wants. The third area of human behavior that economists study is distribution. Once goods and services are produced, they have to be distributed to people. The way goods and services are distributed is crucial to understanding how an economy performs. In the United States, consumer sovereignty drives the production of most goods and services, meaning that consumers are free to decide what they want to purchase in the marketplace. Firms are guided by the market, just as if an invisible hand led them to produce exactly the goods and services that consumers want. No company wants to buy raw materials and pay wages to produce something consumers do not want.

The price of materials and services affects local and international economics. How are prices determined? Price depends on demand and supply. Demand represents peoples' willingness to buy goods and services at different prices. Price is a reflection of how willing people are to buy goods and services. Supply represents the willingness of producers to supply goods and services at different prices. However, supply depends on the time frame being considered to a greater extent than does demand. Producers can better adjust to changes in the market given more time.

130

Economics for Citizenship

A basic understanding of economics is essential if we are to be well-informed citizens. Most of the specific problems of the day have important economic aspects, and we as voters can influence the decisions of our political leaders in coping with these problems. Why is inflation undesirable? What can be done to reduce unemployment? Are existing welfare programs effective and justifiable? Should we continue to subsidize farmers? Does America need to "reindustrialize" to reassert its dominant position in world trade and finance? Because our elected officials largely determine responses to such questions, intelligence at the voting polls requires people to have a basic working knowledge of economics. A sound grasp of economics is literally required knowledge for all good citizens, especially in democracies.

Personal Applications

Economics is also of practical value in business. An understanding of the overall operation of the economic system enables business executives to better formulate policies. Executives who understand the causes and consequences of inflation can make more intelligent business decisions during inflationary periods. More economists are appearing on the payrolls of large corporations, where they gather and interpret economic information used to make rational business decisions. Economics also gives individuals, such as consumers and workers, insights on how to make wiser buying and employment decisions. When the cost of energy suddenly goes up, both workers filling their gas tanks and airline executives have to fit this new cost into their personal and business economics. What should they buy and how much? Which occupations pay well? Which are immune to unemployment? These are all economic decisions.

THE MARKETPLACE ECONOMY

The economy is actually a huge marketplace where people shop for goods and services with the best quality and at the best prices they can find. The economy is like a gigantic machine made up of the following three fundamental parts:

- Supply and demand
- Competition
- Productivity

The study of economics centers on two basic facts: (1) human material wants are virtually unlimited, and (2) economic resources are scarce. The vast majority of the human population (except for saints and ascetics) obeys the first fact. Those who have nice houses and can afford bigger weekend lake houses or wall-to-wall HDTVs, buy them. Humans are acquisitive. The second fact brings into account the effect of limited resources on material wants. Limited resources—oil, finished lumber, polysilicon, nickel, or skilled labor—affect the price and supply of material wants. Economic resources may be classified as the following:

- Property resources like raw materials and capital
- Human resources, such as labor and entrepreneurial ability

Both full employment and full production of available resources requires the prudent administration of scarce resources. However, at any time, a full-employment, full-production economy must sacrifice the output of some types of goods and services to achieve increased

131

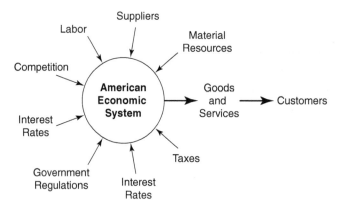

Figure 11.1   The American Economy, An Open System

production of others. As an example, when automotive plants must produce tanks, planes, trucks, jeeps, and so on during war time, the number of automobiles produced is reduced. Also, because resources are not equally productive in all possible uses, shifting resources from one use to another gives rise to the production of additional units of product X, which is in greater demand and brings a higher profit, while sacrificing the production of product Y.

Types of Economic Systems

Historically, the industrially advanced countries of the world have differed esentially on two principles: (1) the ownership of the means of production, and (2) the method by which economic activity is directed. This creates basically three types of economic systems.

A *capitalistic economic system* (see **Figure 11.1**) is characterized by the private ownership of resources and the use of a system of markets and prices to coordinate and direct economic activity. In such a system, individuals are motivated by their own self-interests. The government's role is limited to protecting private property and establishing an appropriate legal framework in which free markets function. The opposite to the capitalistic economy is the *command economy* (communism), which is characterized by public ownership of almost all property resources and the making of economic decisions through a central planning committee. Centralized planning will fail like the dinosaurs who attempted to control a huge body with a small centralized nervous system that could not adapt to rapid changes. A *mixed economy* is a mixture of pure capitalism and the command economy. In this type of economy, individuals are allowed to own some resources, but the government dictates some economic decisions.

Goals of Economic Systems

The various economic systems of the world differ in their ideologies and also in their responses to problems of economics. Critical differences center on a private versus public ownership of resources, and on the use of the market system versus central planning as an economic coordinating mechanism. The following are widely acceptable goals of economic systems:

1. Employment. Suitable jobs should be available for all willing and able to work.
2. Economic efficiency. We want maximum benefits at minimum cost from the limited productive resources available.

3. Price stability. Large upswings or downswings in the general price level, that is, inflation and deflation, should be avoided.
4. Economic freedom. Business executives, workers, and consumers should enjoy a high degree of freedom in their economic activities.
5. Balance of trade. Business executives and workers should seek a reasonable balance in their international trade and financial transactions.

SUPPLY, DEMAND, AND PRICES

The marketplace is where buyers and sellers come together to agree on the exchange of goods and services for money. Producers inform buyers about their goods and services through advertising. The consumer's pocketbook (money) and markets are essential to the American economy because they permit buyers and sellers to influence one another and largely determine what our economy produces, who produces it, how much they produce, and how much the consumer will pay. We come in contact with the phenomena of supply, demand, and prices everyday. For example, supply and demand affect the wages we are paid because wages are in actuality the prices charged for performing work. Wages also affect the prices of goods and services, the prices of raw materials, and the interest paid (price) when we borrow money.

The supply of a product or service is determined by two factors that work together: (1) the cost of producing the product or service, and (2) the selling price. As an example, suppose that buyers are willing to pay significantly more for a product than it costs to produce it. The manufacturer makes a generous profit because everyone has fallen in love with the product. Because of the profit incentive, businesses will usually increase their production to increase their profits. If the profit margin is large enough, other businesses will be stimulated to enter this field and compete for the profits. This competition among sellers normally results in reduced prices or improved products, or both, plus a larger supply of that particular product.

In contrast, if consumers decide not to buy a product, the product is not worth the cost of producing it. When this happens, and if it continues very long, the product will be forced off the market. What is occurring is a balancing act between the supply of goods and services and the demand for them in the marketplace. Though price is not the only determining factor, it plays an important role in this process.

Everyone is familiar with examples of businesses creating demand for products that consumers don't currently want to buy. Dealers may want to clear out the previous year's auto models to make room for new incoming models. The older models don't move until they are reduced in price to stimulate sales. Likewise, department stores hold clearance sales to reduce their inventories as seasons and holidays expire.

COMPETITION

Competition has always been vital to the American economic system. In fact, the vitality of the American economy is based on competition between producers. Those business people who supply the best goods and services at the best prices are generally the most successful. Competition is defined as "a contest between two rivals." Instead of muscles and stamina, companies compete with cost, efficiency, quality, and customer satisfaction. Company A competes with Company B. If Company A's product appears to be inferior or is priced too high in terms of value received, or its service is unsatisfactory, its product will not be competitive and its sales will suffer. Competition, like the forces involved in Darwinian evolution, is one of the factors that causes an economy to constantly evolve and change. Competition for sales can act as a strong stimulus for developing new products and improving current ones (called *innovation*).

Competition is not a factor for monopolies. Monopolies are bad for any country and its people (except the owners of the monopoly). Those who own monopolies really don't care what customers think because they have nowhere else to purchase the product. So a monopoly can maximize profits while treating customers as shabbily as it wants. Monopolies are one reason why black markets thrive. Monopolies can also lead to technological starvation or lack of progress. Why spend money on research and development when they don't have to? Customers have to buy the product no matter how poor the quality or how outdated it is. This is the reason a free competitive economy is good for society and good for technological progress. Companies have to continually enhance their products to make them more desirable than those of their competitors.

Government Regulation of Competition

Because consumers benefit from competition, the United States government has traditionally sought to maintain a competitive environment for business by promulgating laws or regulations intended to prevent abuses in specific areas. Many government regulations also deal with product standards, environmental impacts, and other matters not directly related to competition. Without competition, prices of goods and services tend to be higher than they would be with competition, plus manufacturing output is lower. See **Figure 11.2** and follow the x-axis from left to right. On the left is a monopoly with low or nonexistent innovation and high prices. As competition is introduced across the x-axis, innovation increases and prices come down.

When competition is low, producers can be inefficient without being penalized, and manpower and other resources can be wasted. A lack of competition and rank inefficiency was typical of the manufacturing climate of the former Soviet Union, where workers showed up for work if they felt like it, production lines shut down because parts did not arrive on time, and raw materials and energy were wasted because there was no need to be efficient. No one was penalized because the manufacturer had no competition.

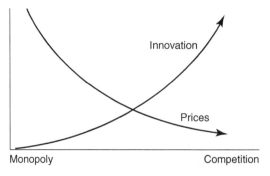

Figure 11.2   The Economic Impact of Competition

Here in the United States, the federal government is an example of a monopoly that is unable to reap the same benefits of new technology, systems, and innovation that private industry does. The government does a poor job of lowering costs, pleasing customers, and just remaining in budget. There are several reasons for poor performance by government organizations, but a primary one is lack of competition. Governments and their agencies have few direct rivals. Amazon.com must outdo all other online booksellers to win readers' money; Google must beat Yahoo. But governments and their agencies are allowed to stagger and stumble along year after year because of a lack of competition.

INNOVATION

Producers of goods and services tend to move to activities where profits will be higher and leave those where profits are lower. This means that workers, facilities, and raw materials shift to new areas of manufacturing constantly. Producers of goods and services always keep in mind **risk**, which is the possibility of loss of value. Producers will not shift assets to a new area of production without a thorough study of the risks involved.

What is **innovation**? The term is often used to refer to new technology, but many innovations are neither new nor involve new technology. The self-service concept of fast food popularized by McDonald's, for instance, involved running a restaurant in a different way, rather than making a technological breakthrough. Innovation is not invention. Novelty of some sort matters. An important distinction is normally made between invention and innovation. Invention is the first occurrence of an idea for a new product or process, whereas innovation is the first attempt to carry it out into practice. Innovation occurs when someone uses an invention—or uses existing tools in a new way—to change how the world works, how people organize themselves, and how they conduct their lives. The test of an innovation is that it creates value.

Probably one of the best examples of an innovative company in America for the year 2008 is Apple. In March 2008, this company was ranked number one in innovation. Led by Steve Jobs, the company has disrupted three different industries—computers (the Mac), music (the iPod), and movies Pixar. (As this book is being revised Apple is preparing to launch the iPhone which should shake up the cell phone industry.) In the organizational context, innovation may be linked to performance and growth through improvements in efficiency, productivity, quality, market share, and so on. All organizations can innovate, including hospitals, universities, and local governments.

The Importance of Innovation

Why is innovation important? From 1980 to 2001, all the net growth in American employment came from firms younger than five years old. Established big firms lost many jobs over those years, and many fell off the Fortune 500 list. Big corporations have been dying off and disappearing from the stock market indexes. Most of the dynamism of the world economy comes from innovative entrepreneurs and a handful of multinationals (General Electric, Boeing, Procter & Gamble, IBM, and 3M, all of which have stayed on the Fortune 500 list more than fifty years) that constantly reinvent themselves.

Innovation adds value (something customers desire) to a product. One of the best examples of innovation is the cell phone. It started out as just a portable phone, then games were added to it, next digital messaging, then the Internet, then the camera phone. Each innovative feature prompted customers to buy the phone with the latest innovative feature. Some pundits place innovation at the pinnacle of modern business, believing that innovation is the lifeblood of any organization. Without it, not only is there no growth, but, inevitably, a slow death. The United States is still the most creative nation in the world. Nearly a third of all patent applications last year came from the United States. However, other countries are catching up fast, especially China, Australia, Korea, and Japan.

PRODUCTIVITY

Productivity is very important to any economic system (see **Figure 11.3**). Productivity describes how efficiently producers (and governments) use their resources—people, facilities, and raw materials. The following are the most important factors influencing productivity:

1. People and their skills, efforts, and motivation
2. Capital resources, their availability, and the efficiency of factories and equipment
3. Technology supplying industrial needs involving new materials, new methods, and advanced processes
4. Organizational effectiveness
5. Government regulations imposing standards or restrictions
6. Working environments that foster both health and work attitudes

Improvement of productivity in business and industry is essential if a country is to maintain competitiveness in selling goods and services both at home and abroad. Productivity levels directly affect inflation and deflation. This is why government and many chief executive officers of large companies closely watch productivity levels. Productivity levels change over the years as economies grow and face new conditions and challenges. Output per

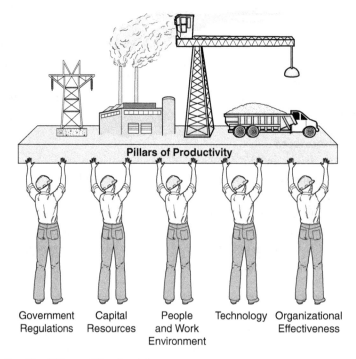

Figure 11.3 The Pillars of Productivity

worker in the United States is usually the world's highest, reflecting a high level of capital investment per worker. But other industrial nations are closing that gap, and this change directly affects our ability to make products that compete in the world marketplace.

As productivity increases, the standard of living also increases. The keys to remaining competitive and increasing productivity are (1) investments in new factories and equipment, (2) improved technology, and (3) a well-educated and trained workforce.

INDUSTRIAL ECONOMICS AND COSTS REDUCTION

People working in the United States today are competing with workers and companies not just in the United States but literally everywhere in the world. To compete successfully, managers, with the help of their workers, must continually improve their goods and services and become more efficient with the use of natural resources and capital goods and services. Managers and workers who successfully improve are rewarded by being steadily employed and receiving consistent pay increases. Managers and workers who do not improve can be faced with the company going out of business or being bought out, which usually results in job cuts and a reduction in pay.

To illustrate this point, let's compare two companies, Cargo and Letcon, that make toy radio-controlled cars. The cost/profit profile for both are shown in **Table 11.1**.

Now, assume that both companies raise wages by 10 percent. However, the Cargo manager told the workers they would receive the raise if they figured out ways to reduce production costs by 10 percent. That would not require an increase in the product price and would keep their prices competitive. The workers succeeded in reducing waste by 10 percent and received their raises. The managment at Letcon just raised everyone's wages and then raised its toy price to pay for it. At first, everyone at Letcon seemed to get the better deal because they received the same pay raise as Cargo and did not have to put out the extra work that workers at Cargo had. Who got the better deal? **Table 11.2** illustrates the two companies' profile sheets one year later.

Table 11.1   Cost/Profit Before Wage Increase

	Cargo	Letcon
Total toys produced (per year)	1,000,000	1,000,000
Income		
Price per toy	90¢	90¢
Total income (year)	$900,000	$900,000
Annual cost		
Raw materials	$100,000	$100,000
Wages	$300,000	$300,000
Fixed costs	$300,000	$300,000
Total	$700,000	$700,000
Gross profit	$200,000	$200,000
Taxes	$ 50,000	$ 50,000
Net profit	$ 150,000	$ 150,000

Table 11.2 Cost/Profit After Wage Increase

	Cargo	Letcon
Total toys produced (per year)	1,000,000	1,000,000
Income		
Price per toy	0.90¢	0.99¢
Total Income (year)	$900,000	$495,000
Annual cost		
Raw materials	$100,000	$100,000
Wages	$330,000	$330,000
Fixed costs	$300,000	$300,000
Total	$730,000	$730,000
Gross profit	$900,000	$495,000
Taxes	$ 50,000	—
Net profit	$ 120,000	−$235,000

Now it is easier to see that Letcon's good deal is no longer as good as it first seemed. As soon as Letcon raised its prices, quite a few consumers stopped buying its toy cars, and Letcon lost half its business. Unless something is done, Letcon will have to shut down because it will soon be bankrupt. Thus, all the Letcon workers could lose their jobs. Cargo workers, however, still have their jobs with the higher wage rates. Also, if Letcon goes out of business, Cargo stands to profit further through expansion by picking up Letcon's customers.

This story has two morals. The first is that there is no free lunch (free wage increases). Someplace, somewhere, someone is going to pay for the lunch. In Letcon's case, lunch was paid for by laying off workers because of a drop in profits. The second moral is that the welfare of the workers and the company interlock, and each depends on the other. To believe that wage raises can be shoved through and paid for by increasing the price of goods and services is not realistic if competitors don't raise their prices. All producers—management and labor—must work together to gain wage increases through productivity gains or the company's long-term survival is at risk.

THE GLOBAL MARKETPLACE

In the last twenty years, economies have become globalized. **Globalization** is the integration of national economies into the international economy through trade, foreign direct investment, capital flows, migration, and spread of technology. This literally means that consumers can purchase anything from anywhere and sell it anywhere. Consumers do not have to buy locally, regionally, or nationally if they do not want to. One of the biggest changes that has taken place in the last twenty years is the proliferation of information. People in highly developed countries receive information directly from the source through TV, the Internet, cell phones, and other mass media (regional and national magazines and newspapers). For business purposes today, every place is the same as every other place.

Wherever people go in the world, they see advertisements for Sony, IBM, and Nike. News footage shows villagers in Nepal wearing T-shirts with faces of American country-and-western singers on them or villagers deep in the Amazon wearing Pittsburgh Steelers ball caps. Producers distribute their products to customers all over the world as they seek to grow their customer base.

Today, people have the information available to know where to buy the least expensive or best quality product. Information, the computer, and transportation advances created the global market, not just in manufacturing but also in services. Think of a hospital in the United States with a reputation for delivering excellent services and saving many lives when other doctors and hospitals had considered the patient as good as dead. Because of information technology today, hospitals compete worldwide for wealthy Arab or German heart patients because they can advertise their message, they can access and communicate with those customers, and speedy transportation makes getting to their hospital easy for people.

The United States is not the only country marketing and selling medical services. Because U.S. services for many serious operations are so expensive, many Americans now seek surgery in special hospitals in India and other Asian nations that offer the same surgery by highly skilled surgeons for one-third the price.

Every country is being directly or indirectly shaped by globalization. It is not a historical accident that East Germany, the Soviet Union, Brazilian state-owned industries, Chinese communism, General Motors, and IBM all either collapsed or were forced to radically restructure at roughly the same time during the 1990s. Globalism has forced companies to face the following four key elements in a modern economy:

- Quality has replaced quantity as the way to measure goods and services.
- People (human talent) matter most, not a country's physical resources.
- Quality improvements are being made in some industries at a revolutionary rate.
- The economy is now customer-driven, as opposed to "the customer be damned" days of old.

Global Consumers

Consumers have caused political revolutions all over the world. The falling away of the boundaries between Eastern Europe and Western Europe is all because people in Eastern Europe wanted to shop but couldn't. They saw all the mass-produced goods—cars, microwaves, computers, color TVs—on television and in other forms of mass media, and they wanted to become consumers. There is even speculation that earlier Russian disarmament agreements were caused by Bon Jovi and Levi Strauss, not the pressure of NATO or American military superiority. Meaning, the Russian people began to dismantle their military hardware because they were aware that their lifestyle was not on a par with the rest of the world. They didn't want more tanks; they wanted Nikes, Panasonics, and Wranglers. Ideology, how to be good—but materially poor—Soviet citizens, ceased to be a central issue. Life was short and people were looking for the good life. Power has passed to the consumers.

All over the world now, customers have more power. Systematically, in industry after industry, power is shifting from the people who sell to the people who buy because:

- Consumers have more choices (styles, models, colors, special features).
- They can get products from anywhere (China, Germany, France, India, and so on).
- More companies (worldwide) are competing for their attention and money.

More competition, more choices, and more information put more power in the hands of consumers, and that, of course, drives the need for quality. In 1980, Mercedes had just six competitors in the arena of luxury automobiles. In 2008, it had more than twenty-five competitors.

GLOBAL COMPETITION

Two hundred years ago in the United States, most trading was conducted locally. Then, about a hundred years ago, the United States became a national economy, and a region of the world economy, just as Europe did. At about the same time, Japan came out of three hundred years of isolation and began building a modern industrial economy to compete with the West. Over the last fifty years, as transportation and communication costs declined, the world economy became globalized. Most large American, European, and Japanese companies have investments all over the world. The United States has billions invested in Europe, and the countries of the European Union have billions invested in the United States. American companies are making large investments in China, India, Ireland, and Vietnam. India and China are making investments in the United States.

Globalization and IBM

In June 2006, IBM held its annual investors day in the grounds of the Bangalore Palace, India's equivalent of Silicon Valley. It was held to impress the analysts and investors who turned up to hear speeches by the president of India, the country's leading telecom entrepreneur. The annual investors day is usually held in New York but IBM sent a strong message by going to Bangalore. With 53,000 IBM employees, India is now at the core of IBM's strategy. Samuel J. Palmisano, IBM's CEO, pointed out to his investors in Bangalore that the Indian market had become one of the fastest growing in the world for IBM, with revenues rising by 40 percent to 50 percent a year. IBM now has more employees in India than any other country in the world except the United States.

IBM's Indian venture highlights that emerging economies increasingly count as a threat, as well as opportunity, to established global firms. IBM intended to shape its strategy, management, and operations as a single global entity. It will put people and jobs any place in the world where the costs are right, the skills are right, and the business environment is ripe. In this approach, work flows to the place where it will be done the most efficiently and at the highest quality. The forces behind this move, he said, were irresistible. Places like India, he said, do not just mean cheap labor. India has a skilled workforce and a friendly business environment. There is an important message here for workers in all American companies and it reflects what we have learned in Chapters 7 and 10—companies cannot survive or remain in a country where the workforce does not have technical expertise and talent.

Can We Un-Globalize?

Some Americans and members of Congress want to protect American companies or jobs by keeping foreign goods and factories out. In the 1980s, some Americans claimed they would not own a Japanese car or any foreign car because it was unpatriotic and took away American jobs. "Buy America" campaigns started up with the goal of persuading American consumers to buy American products to help reduce the trade deficit. Ultimately, however, American consumers were driven by quality.

Today, buying American is more difficult than it sounds. How do you define an *American* car? Is it a Honda assembled in Ohio by American men and women, or is it a Pontiac LeMans assembled at a highly automated Korean plant, based on a design by Opel in Germany, then sold under an American name? In the auto industry, American and foreign firms now routinely engage in joint ventures and are suppliers to, and customers of, one another. American cars are more than likely international products, not purely American products.

Though globalization is disruptive at first, it is irresitible. There is no going back to pre-globalization. Any country attempting to do that would be like a nation deciding it wanted to return to the horse-and-buggy days.

SUMMARY

Economics is the production, distribution, and consumption of goods and services. A basic understanding of economics is essential to being well-informed citizens. Most of the specific problems of the day have important economic aspects. An understanding of the overall operation of the economic system enables business executives to better formulate policies. A nation's economy is just a marketplace where people shop for goods and services with the best quality at the best prices they can find.

Competition has always been vital to the American economic system for creating the best goods and prices. The vitality of the American economy is based on competition between producers. Producers rely on productivity to remain in business. Productivity describes how efficiently producers use their resources—people, facilities, and raw materials. Producers must consistently increase productivity to remain competitive in the global market.

Today, customers exert tremendous power over producers. Systematically, in industry after industry, power is shifting from the people who sell to the people who buy because more companies are competing for their attention. Globalization has increased the power of consumers.

REVIEW QUESTIONS

1. Explain why economic knowledge is important to good citizenship.

2. List five goals of economic systems.

3. A person's willingness to buy goods and services at different prices is called _____.

4. List four economic issues that all national leaders face.

5. Discuss the relationship of supply, demand, and price.

6. Discuss the two factors that work together to determine the supply of a product or service.

7. Explain how competition drives an economic system.

8. Describe two ways in which monopolies are bad for a society.

9. List six factors that influence productivity.

10. Explain the following statement: "The welfare of the worker and the company depend on each other."

11. Discuss how information technology has affected the global marketplace.

12. Explain why customers have more "power" today and how that power influences companies.

13. Define the following: *economics, inflation, risk, standard of living.*

GROUP ACTIVITIES
Read the following articles and be prepared to discuss them.

The Razor Wasn't Built in a Day

Article written by Mike Speegle, 2008
Prior to the invention of the safety razor, men shaved with the single-edged knifelike strop razor often seen in Western movies. It took seventy years from the invention of the safety razor by King Gillette in 1901 to the launch of the company's two-blade version in the 1970s. Since then, the evolution of the safety razor has not changed much. The razor went from one blade to two blades, and then three blades became the world's top seller when launched as the Mach3 Turbo in 1998. Gillette's big rival, the Wilkinson Sword®, went to four blades with the Quattro® in 2003. And sure enough, Gillette jumped to five blades with the Fusion® in 2005. The Fusion may be the limit of genuine innovation in safety razors.

Competition

By Mike Speegle, March 2008
Competition in today's business world can best be described by an old Swahili proverb from Kenya: "Every morning in Africa a gazelle wakes up. It knows it must run faster than the lion or it will not survive. Every morning in Africa a lion wakes up and it knows it must run faster than the slowest gazelle or it will starve. It doesn't matter if you are the lion or the gazelle, when the sun comes up, you better be running."

The publishing tycoon Rupert Murdoch captured the essence of this proverb when he said that, "The world is changing very fast. Big will not beat small anymore. It will be the fast beating the slow."

In today's fiercely competitive business and economic climate, we are all running. And most of us—like the gazelle in the proverb—are looking over our shoulders while we run. Many factors determine which countries and companies run faster than the others and which ones are left behind. Two factors that are most important are *innovation* and *entrepreneurship.*

Steve Jobs, the CEO of Apple Inc., saw his personal computer field losing major market share years ago. Jobs and his company executives quickly branched into a different but related field to change the way that the world listens to music by launching the iPod portable music player, and then the innovative iPhone. Twelve years ago, portable music came in the form of a bulky boom box cassette player that ran on a dozen AA batteries, or a somewhat less bulky Walkman with large, heavy earphones. Today, a sleek, attractively designed iPod scarcely larger than a thick credit card emits high-quality sound through ultralight earbuds. And, because the iPod requires a computer to download music, the product rejuvenated Apple's core personal computer business. That was innovation in practice.

The economic competitiveness of the United States and how it stacks up against the countries of the European Union (EU) and of East Asia were discussed in the 2006 *Lisbon Review.* The results measured EU national economies for economic liberalization, innovation, research and development, regulatory environment, financial services, and social inclusion. Denmark came out on top as the EU's most competitive, dynamic, and innovative economy. Finland and Sweden were right behind Denmark in the rankings, the Netherlands was rated fourth, Germany number five, and the UK finished in the sixth position.

The part of the review that compared the EU, American, and East Asian economies gave Americans reason for both pause and for optimism. Even though the U.S. economy surpassed the collective EU and East Asian performances by a fairly comfortable margin, the EU's top six performing countries outranked the United States. But the United States outperforms all EU member states and East Asian nations in innovation, and research and development, which drive the productivity and prosperity of modern economies. The capacity for innovation is a key competitive factor in the globalized age.

The second factor that determines winners and losers is the entrepreneurial spirit. In America, the role of the entrepreneur in society and the relationship of the entrepreneur to our culture are unique: the entrepreneur is king. An *entrepreneur* is a person who organizes, manages, and assumes the risk of a business. With twenty straight quarters of economic growth, 5.7 million new jobs created within a three-year period, two-thirds of which were created by entrepreneurs, something encourages entrepreneurs in America. In Denmark in 2005, an American businessman visited with a group of talented student body leaders at University of Aarhus. During their discussion, they described their feelings toward America. Essentially, they said that America was a land of opportunity, a land where hard work could lead to success and the American dream. If that is true, what are the factors that make it exactly that? The following are two reasons why America is still the "land of opportunity."

First, the primary role of government in America is to foster innovation and entrepreneurship. Since the latter part of the twentieth century, American presidents have agreed that the highest and best purpose of the federal government is to foster the spirit of innovation and entrepreneurship in American people. Ronald Reagan was the first American president to specifically identify this role of government when he said, "It is not government's, but it is the entrepreneur and their small businesses who are responsible for almost all the economic growth in the U.S." In Europe and much of the rest of the world, the historic and traditional role of governments has been quite different.

Second, America is the land of the entrepreneur because America rewards those who take risks. Americans celebrate the attempt more than punish the failure. The most famous American examples of this are Thomas Edison and Orville and Wilbur Wright. Edison came up with more than nine hundred ways not to invent a lightbulb. The Wright Brothers failed more than a hundred times to get their airplane off the ground. But, they finally succeeded. Venture capitalists in Silicon Valley will tell you that getting funding is easier if you have tried and failed, than if you have never tried before. Business failure is not the end of a person's career, but rather just another chapter in an individual's professional development.

CHAPTER 12

Quality as a System

Learning Objectives

After completing this chapter, you will be able to:

- *Write a definition of a system.*

- *Explain what is meant by the phrase, "Quality occurs in a system."*

- *Describe the difference between a customer and a supplier.*

- *List five subsystems associated with a styrene process unit.*

- *Explain why it is important that all subsystems work properly.*

- *Explain how external customers determine value-added costs.*

- *Explain the difference between horizontal alignment and vertical alignment.*

INTRODUCTION

All work consists of a series of steps, and these steps constitute a system. What is meant by "quality being a system?" What is a system? And what does a system have to do with any business or with the processing industries and plant operations? A system is a way of looking at a group of things or steps that helps us to understand how these things work together. A system can take in a lot of equipment and multiple processes (the big picture) or a few pieces of equipment (a snapshot). We will define a *system* as a *collection of different equipment and processes that are interrelated.*

To add clarification to the definition, a **system** is (1) a group of equipment that works together to accomplish a task and/or (2) a collection of processes dependent upon one another to complete a task or product. We are interested in the second definition because it is the big picture, meaning it includes everything. Systems are how work gets done.

SYSTEMS AND SUBSYSTEMS

This second definition of system probably would be easier to understand if we look at an example of systems. Consider the automobile. What makes up an automobile? It has cylinders, pistons, fuel injectors, a timing mechanism, air and fuel filters, a transmission, a drive train, a battery, tires, and many such other items. Sure, all that stuff makes up an automobile, but it doesn't help you to understand much about how each contributes to a properly functioning automobile.

To function properly, a car must start, get us to where we want to go, return us back home, keep us safe and comfortable while in the car, and give us an enjoyable ride. Ford, Toyota, GM, and Honda use the system criteria when they design new models or improve the current models for next year's sales. They break down a car into subsystems, such as the drive train, emission control system, electrical system, braking system, lighting system, and so on. They look at each subsystem separately and the equipment that makes up that system and determine how the equipment grouped in each subsystem can be improved while also reducing the cost of the system. Before they decide to improve a subsystem, they must understand its relationship to other subsystems and ensure that the improvement will not adversely affect other subsystems. Several years ago, one automobile manufacturer decided to digitize everything it could in its newest model. The car was literally going to be a computer on wheels. Unfortunately, problems in one system can affect the operation of another system. This company's digitization improvement caused the new model to act like a computer in that it froze up (stopped running) at various times or frequently would not start.

Key Ideas about Systems

A system (also called a *process*) is an organized group of related activities (subsystems) that work together to transform one or more kinds of input into outputs that are of value to customers (internal or external). This definition communicates several key ideas, such as:

- A system is a group of activities.
- The activities that make up a system are related and organized.
- All the activities in a system must work together toward a common goal.
- Systems exist to create results for internal and external customers.

A system can be viewed as a value chain in which each activity or step contributes to the end result. Some activities directly contribute value, whereas others may not. All activities consume enterprise resources, however. The challenge for managers is to eliminate steps that do not add value and to improve the efficiency of those that do.

THE PROCESS UNIT AS A SYSTEM

The concept of a system can be very flexible. You can make a system out of anything as long as you can relate the parts that make up a system to each other. Consider a petrochemical plant (a plant that makes chemicals), specifically, a styrene processing unit. This unit makes styrene monomer, a water-white chemical that is a basic building block for

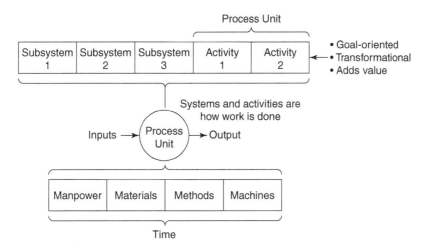

Figure 12.1   Process Unit System and Subsystems

insulating materials, paints, automobile tires, and so on. Assume the unit was built twenty-five years ago for a cost of twenty-two million dollars. Why was the unit built—to make styrene? No. Every investment in a business venture has one primary goal—to make a profit. This styrene unit has many competitors, all selling styrene monomer at the market price. If all the styrene manufacturers sell their product at the same price, their profit will come from the efficiency of their processes. In turn, the efficiency of their processes will be determined by their systems, which are a combination of subsystems that are made up of manpower, methods, materials, and machines. All of these comprise the styrene manufacturing system (see **Figure 12.1**).

The styrene unit is not a stand-alone unit, meaning it depends on services or supplies from other units and departments within the plant, which are subsystems involved in the overall styrene manufacturing system. The following are examples of subsystems involved in the styrene processing system:

- Utility systems—plant air, instrument air, electricity, natural gas, steam, and so on
- Auxiliary systems—hot oil, flare, cooling water, refrigeration, and so on
- Raw materials (feed, catalysts, additives)
- Process unit subsystems (reaction, separation, finishing sections)
- Quality control
- Marketing
- Human resources (payroll, vacation, insurance, and so on)
- Shipping (bills of lading, railcars, barges, tank trucks)
- Product storage (run-down tanks, tank farm)
- Health, environmental, safety, and security

The System and Profits

The process technicians and unit engineers either cannot make on-specification styrene monomer product (1) economically or (2) sell it and make a profit, unless all of the subsystems operate correctly. The unit cannot function unless the utilities and auxiliary systems operate correctly. Also, making on-specification styrene is pointless if the company cannot sell it because the marketing department is ineffective, the shipping department ships product late and enrages customers so that they choose a new supplier, or the plant receives a large monetary penalty from the Occupational Safety and Health Administration (OSHA) or Environmental Protection Agency (EPA) because of safety or environmental releases and therefore reduces available capital for the site. Every subsystem must work together and work correctly for unit efficiency and effectiveness to produce both a desirable product and a profit.

The most efficient styrene process unit is at the mercy of its subsystems. Like the human body's subsystems (heart, liver, lungs, muscles, nervous system), all the subsystems that make up the styrene process unit system must work together correctly for the process unit to survive. Can a process unit die? If the styrene unit becomes unprofitable, it will be sold to another company that will operate it at lower wages and benefits, or it will shut down and sell its equipment to China or India. To the parent company, it is "dead." Thus, the styrene process unit is dependent on its internal customers and suppliers that make up its subsystems.

The quality of the styrene monomer product for sale is dependent upon all the styrene unit subsystems. This does not mean only the quality of the on-specification styrene monomer product, but the quality of the styrene manufacturing system that is responsible for:

- Producing on-specification product
- On-schedule production
- The efficiency of the system that makes the product to be sold at market price while still making a reasonable profit
- The marketing and transportation system that creates demand and delivers the product on time to the customer

All of these together make up a styrene process system. Remove any of the subsystems of internal customers or suppliers, and the big system either becomes crippled (inefficient) or ceases to work. This is why communication, teamwork, and interpersonal skills are so essential in the processing industry, or in any industry for that matter.

CUSTOMERS AND SUPPLIERS OF THE SYSTEM

Besides equipment, a system is made up of customers and suppliers. Although process technicians may not have direct contact with external customers, they each have internal customers for the products and services they produce. Improving these internal customer-supplier relationships is absolutely vital to keeping the ultimate customer—the external customer—satisfied, and for continuous improvement. Becoming customer-focused starts with identifying the key customers (both internal and external) and clarifying their requirements. For operations to be truly effective, organizations should monitor the requirements of customers and the behavior of suppliers. Supply chain management is a critical success factor.

The Customer-Supplier Chain

To improve a process (system), employees must begin by creating a **line of sight** to the external customer. In other words, they must understand where they fit in and how they contribute to the chain of customers and suppliers in their organization. The customer-supplier

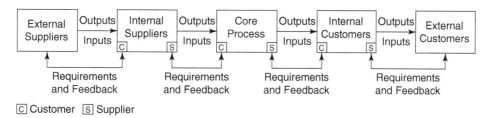

Figure 12.2   Internal Customer-Supplier Chain

chain in **Figure 12.2** illustrates that all work is a process in which customers and suppliers are linked to one another. Some of these customers are external to the organization; others are internal. Figure 12.2 describes the relationships between supplier inputs, the process (work) performed, and outputs provided to customers in the production chain.

The customer-supplier chain is present in every work process, whether the process involves purchasing, shipping, production, maintenance, or marketing products. The concept of the customer-supplier chain is important enough to study in more detail.

First, all work is a process that requires inputs of materials, energy, and labor. Assume one of the inputs is crude oil at a refinery. Value can be added to the crude oil by converting it into gasoline and diesel fuel. Adding value is going to require more labor, materials, and energy, but results in the output of the process (gasoline and diesel fuel). If an employee is involved in this process, the work she does is part of this customer-supplier chain and is not an isolated activity. The quality of that work affects the process (system).

That employee is a supplier in this process when her outputs are inputs to others in the refinery or to external customers. As an example, an operator is a supplier to the quality control laboratory when he submits samples to the lab. Then, he becomes a customer when he receives analytical results from the laboratory. On a single day, the varied tasks he performs may alternate his roles of supplier and customer several times.

Value-added activities, also called value-added work, are those activities that are necessary to turn inputs into outputs that will meet or exceed customer requirements. The objective is to carry out these activities in the most efficient manner possible. Employees are both customers and suppliers. They will likely have many more internal customers and suppliers than external ones. By providing outstanding service to internal customers, they enable the unit to provide outstanding service to external customers. When workers know and understand the requirements of customers in the chain, the process is aligned and can achieve its quality goals. All workers in the customer-supplier chain must understand their roles and do their work correctly if their process is to be an efficient and profitable process.

Maintaining the line of sight to external customers is important because ultimately only external customers pay revenue to the company. Internal customers generate additional costs. External customers determine whether those costs are truly value-added by their decisions to buy or not buy the products and services at the asking price. In other words, if a process requires a lot of unnecessary work for the final product because the process is very inefficient, customers will not pay the price of the product because all that unnecessary work has made it more expensive than a competitor's product. Once employees understand their lines of sight to the customer and how their work is accomplished, they can analyze their work to identify opportunities to eliminate waste and improve the process.

External customers who buy the finished product are the *key customers* and represent the highest priority. Employees may be inclined to consider the management team, or whoever signs the paycheck, as a customer, but that is not true. Their functions exist for a real business reason, not to please those who sign their paychecks. They are almost certain to please the management when they delight internal customers with an error-free process. The result is good product at low cost.

Alignment

As stated in the previous explanation of the customer-supplier chain, when the requirements of every customer in the chain are understood and being met, there is alignment. This type of alignment, called **horizontal alignment**, extends from the ultimate, external customer, all the way back through the organization to external suppliers. This horizontal alignment is like a flowchart that shows all the steps required in the process to put the finished product in customers' hands. Horizontal alignment allows companies to determine customer needs and continually improve the ability to meet those needs.

Another kind of alignment is equally important to work teams and is often called **vertical alignment**. It extends from organizational goals and strategies down to individual and team performance. Vertical line of sight provides important information about how the organization chooses to spend resources to improve its overall performance and capabilities. Because amounts of people, money, and time are always limited, goals and strategies provide the information required to prioritize alternative courses of action. Vertical alignment ensures that a team isn't working on improvement of something that the strategy has determined is unimportant. By developing a thorough understanding of organizational goals and strategies (vertical line of sight), a company will be able to prioritize alternative improvement opportunities. Most of these possible improvement opportunities will be identified as the result of developing a thorough understanding of customer needs (horizontal line of sight). For this reason, a combination of horizontal and vertical alignment enables a team to ensure it is focusing its improvement efforts and is doing the right things right.

REQUIREMENTS (THE WHAT QUESTIONS)

Once process unit supervisors or operators have identified their customers, their next step is to develop a clear understanding of their customers' expectations and translate these into agreed-upon requirements. The following are basic questions to ask:

1. WHAT do you need from me?
2. WHAT will you do with what I give you?
3. Do you understand WHAT I can give you?
4. Do we agree on WHAT the requirements are?

What do customers need? There is no easy answer to this question because customer satisfaction is often based more on the customers' sometimes fuzzy expectations than on outright needs. This is why it is important to get a clear understanding of requirements that customers and suppliers agree upon. Often, customers have a set of expectations, some that are consciously stated wants and needs, and some that are not stated and must be uncovered through probing questions or research. Through discussion, wants and needs can be clarified, and previously unrecognized needs may even be pointed out to the customers (maintenance department, laboratory, technical service). New expectations will continue to emerge over time. For this reason, the job of clarifying customer requirements is never finished.

A True Customer-Supplier Story

I had an interesting experience relating to internal customers when I was a quality control laboratory supervisor at a petrochemical plant. The experience clarified the requirements by asking, "Do you understand WHAT I can give you?"

One large process unit occasionally experienced problems and sent in samples to the quality control laboratory for gas chromatography (GC) analysis to aid in troubleshooting the upset process. Process unit supervisors and operators complained about the slow response of the laboratory and how the slowness of that response cost the process unit money. Members of the laboratory group and the process group each thought the others complained about anything and everything.

One day, the process unit experienced another upset, and an operator personally brought two samples to the lab for immediate gas chromatography analysis. The operator and I went into the GC lab, and I asked if the required GC was available and was told it would take about forty minutes because a sample had just been injected into the GC. The operator demanded that the sample be removed and his sample be analyzed. I explained to the operator that a sample required about forty minutes to pass through the GC column and out into the detector. I explained how the GC worked, and I could see understanding click on in the operator's eyes. The operator called his supervisor so he could hear what he had just been told. When operations understood what the lab could give them, the lab and process unit created a process to analyze their samples more rapidly. Later, the process unit also bought a new GC for the laboratory, dedicated only to their samples.

Note: This incident happened before that petrochemical site adopted and began training all employees in Philip Crosby's quality improvement process.

Internal Customers

Instead of defining customers as those who purchase commodities or services, let's instead call a customer anyone—an individual or an organization—that receives and uses what an individual or an organization provides. Based on this definition, customers are no longer those outside the bounds of the organization. An entirely new category of customers, Ishikawa's internal customers, has emerged. These customers can be inside the organization supplying a commodity or service.

Consider a simple example in a manufacturing environment, specifically a chemical plant that produces styrene. A sister unit in the plant sends some benzene residues to the styrene unit as a supplemental feed. Three maintenance technicians are assigned to the styrene unit to keep its equipment running and perform preventive and predictive maintenance. The analyzer group has assigned an analyzer technician to be responsible for maintaining the

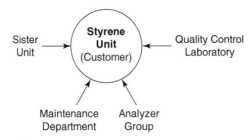

Figure 12.3   Styrene Unit as Customer

calibration of the analyzers on the styrene unit. The quality control laboratory analyzes samples 24/7/365 for the styrene unit. Thus three work groups (see **Figure 12.3**) in the chemical plant—the sister unit, maintenance technicians, and analyzer technicians—provide a product or service to the styrene unit. The styrene unit is the customer of these three work groups.

What are the implications of this new customer definition, and the concept of an internal customer? Consider the following questions:

1. Does the sister unit supplying benzene residue have to meet the specifications and expectations of the styrene unit? What happens if they don't?
2. Do the maintenance technicians have to meet the specifications and expectations of the styrene unit? What happens if they don't?
3. Do the analyzer technicians have to meet the specifications and expectations of the styrene unit? What happens if they don't?
4. Does the quality control laboratory have to meet the specifications and expectations of the styrene unit? What happens if they don't?

Hopefully, you now recognize the interdependence of a process on its internal customers and suppliers. All are important.

QUALITY AS A SYSTEM

Only through a systemic approach to quality can continuous improvements be made and prevention implemented. The section of this chapter titled "The Process Unit as a System" lists ten subsystems that support the styrene process unit. Each subsystem (supplier) must know the requirements (specifications) that they are required to supply twenty-four hours a day, seven days a week, 365 days a year. In addition, all four crews (twenty operators) of the styrene unit must know and understand their jobs and perform them correctly, plus know and verify on a daily basis that the subsystems meet the unit's specifications. The process unit is a huge system, and the quality metrics of the styrene unit are dependent on the knowledge and skills of the operators in understanding and operating this system.

The quality metrics of the styrene unit are not something as simple as (1) meeting the production schedule and (2) making only on-specification styrene. Those are not quality metrics; they are essentially the minimum requirements to remain in business. However, they do not guarantee that unit will remain in business. Remember, other companies competing for customers also have styrene units. The quality metrics for the styrene unit might consist of:

1. Meeting the production schedule
2. Making only on-specification product
3. Reducing waste equivalent to 7 percent of the unit's annual budget
4. Increasing productivity by 2 percent this year
5. Ensuring no OSHA-recordable injuries
6. Creating prevention measures for trouble areas
7. Mandating that all crew members attend sixteen hours of troubleshooting techniques this year
8. Assigning members of each crew to work on and eliminate a specific unit internal failure that has occurred more than three times

From this list, it should be obvious that an operator who shows up for work on time and does his daily operating duties is only doing the minimum, and doing the minimum is an excellent way for the unit to go out of business in a few years. Only through a systemic approach will the styrene unit operators be able to accomplish all of these unit goals.

As an example, the styrene unit is buying steam from the Number 1 boiler unit at 78¢ a pound. Finding and replacing lost or loose piping insulation will lower the steam usage. Detecting and repairing nonfunctioning steam traps will save more money. These savings contribute to goal number three. If operators list how many defective types of valves have been replaced in nine months, how they failed, and the manufacturer's name, they might switch to a more reliable valve, thus reducing maintenance costs. Operators who only show up for work on time and do the minimum to keep from getting fired will likely not make these improvements.

Only operators who understand the interdependence of the process unit equipment and the subsystems that support the process unit can contribute to continuous improvement, thereby supporting the survival of the process unit. This is the systemic approach to quality.

SUMMARY
All work consists of a series of steps, and these steps constitute a system. A system is (1) a group of equipment that works together to accomplish a task and/or (2) a collection of processes dependent upon one another to complete a task or product. A system can be viewed as a value chain in which each activity or step contributes to the end result. Some activities directly contribute value, whereas others may not. All activities consume enterprise resources, however. The challenge for managers is to eliminate activities that do not add value and to improve the efficiency of those that do.

To improve a process (system), employees must begin by creating a line of sight to the external customer. In other words, they must understand where they fit in and how they contribute to the chain of customers and suppliers in the organization. Once the process unit supervisor or operator has identified her customers, the next step is to develop a clear understanding of their customers' expectations and translate these into agreed-upon requirements.

A process unit in a petrochemical plant or refinery can be a huge system composed of many subsystems, and the quality metrics of the process unit is dependent on the knowledge and skills of the operators who understand and operate this system.

REVIEW QUESTIONS

1. Write a definition of a system.

2. Explain what is meant by the phrase, "Quality occurs in a system."

3. Describe the difference between a customer and a supplier.

4. List five subsystems associated with a styrene process unit.

5. Explain why it is important that all subsystems work properly.

6. List five quality metrics for a process unit.

7. Explain how external customers determine value-added costs.

8. Explain the difference between horizontal alignment and vertical alignment.

9. Define value-added work.

10. Ultimately, only _____ pay revenue to a company.

GROUP ACTIVITIES

The following two articles reveal how individuals or groups helped a company compete or survive. They reveal how reducing waste and making several very small changes (incremental changes) that add to major savings are possible. Read the two articles and answer the questions associated with each article, then prepare to discuss the articles.

Incremental Changes for One Airline

By Mike Speegle, March 2007

Working the midnight shift in late 2004, a dispatcher for one of the largest U.S. carriers monitored planes flying from Dallas to Colombia, South America. He noticed that some burned far more fuel than others, and wondered why. The weather patterns were usually the same, so he knew that couldn't be it. He also knew it didn't matter which captain was flying the plane. The dispatcher, curious, kept wondering why there was such a fuel discrepancy. The dispatcher, whose interest in engine performance came from his hobby of drag racing, compiled a list of gas-guzzling jets and gave it to his superiors. Officials at the airline believed they could save $11.2 million a year by fixing the worst offenders in the 726-plane fleet. Some of the fixes were simple, sometimes requiring small mechanical adjustments; others required smoothing out small dings that affect a plane's aerodynamics and create increased friction. This was part of a larger plan to save $117 million in fuel a year. Although this amount was not much of a savings for an airline that lost several billion since 2001, the company management said they had to start somewhere.

Jet fuel accounts for nearly 20 percent of the airline's costs, its second-largest expense behind labor. The airline spent $2.8 billion on fuel the previous year, and, with stubbornly high crude oil prices, estimated the 2005 cost to be $3.5 billion. The increased cost of fuel turned what could have been the first profitable year since 2000, into another money-losing year. This was the same story as those at many other airlines. The Air Transport Association estimated that U.S. carriers would pay $6 billion more for fuel by 2005, because of higher crude oil prices. The airline cited fuel when it tried to raise fares but backed off the fare increase and continued to lose money when some rivals declined to boost their prices.

Among the fuel-saving steps taken by the airline was to have airplanes taxi on one engine instead of two, and shut down auxiliary gas-powered engines while planes were parked at the terminal. To reduce weight and improve mileage, the planes had been carrying half as much reserve fuel on international routes after approval by the Federal Aviation Administration. The airline also began using its planes as tankers, ferrying fuel from places where fuel is cheap instead of filling up where it is costly. When jet fuel cost $1.58 a gallon in California but $1.16 in Dallas the previous summer, the airline saved $400 a flight by filling up in Texas. Ferrying fuel this way saved $15 million that year.

Computers are used to calculate how to spread the weight of passengers when flights aren't full. An airliner with an uneven load could cause the plane's nose to rise, increasing drag and reducing mileage. Changing the center of gravity on each large airliner by just eleven inches would save $5 million a year. Management knew these weren't huge savings—a $100 million or more—but knew it could find lots of smaller ones that add up to huge savings. The airline was considering reducing the maximum cruise speed of its jets to conserve fuel and installed "winglets" on its planes, like Dallas-based Southwest Airlines. Southwest began installing the upturned wingtips on its Boeing 737-700s and reported that they boost mileage 3 percent to 4 percent, saving about $9 million. However, the winglets cost $725,000 a pair, so recouping its investment could take a while for the airline.

Management at the struggling airline said they were going to heal the wound and stop the bleeding (loss of profitability and threat of bankruptcy). They were going to keep making incremental changes and create a culture of change. Even if fuel prices dropped lower, they intended to keep finding ways to work smarter. They admitted they should have been doing this years ago.

1. Why not raise the cost of a ticket to cover the increased fuel costs?

2. Why not charge passengers who are overweight more money?

3. Why not offer special incentives to people who are not overweight?

4. Why didn't management foresee and plan for rising fuel prices?

5. Do you understand why stewardesses on half empty planes would ask passengers to remain in their assigned seats?

6. An airline makes money only when its planes are in the air carrying passengers. How can ground employees (baggage handlers, ticket agents, refuelers, maintenance personnel) help the airline toward profitability?

Productivity Pays Off

By Mike Speegle, April 2007

In 2006, assembly line workers at trim station #1 at the Mercedes-Benz factory in Alabama noticed they were laboring a bit too hard. To reach fasteners and bolts needed to install consoles and wiring harnesses in the Mercedes M-Class SUVs, workers had to "slant walk" zigzag-style to parts bins a few feet down the line. Each trip required five or six steps, and workers sometimes bumped into one another. Management could have responded to such a complaint with, "Shut up and get back to work," but when a supervisor on one of the plant's continuous improvement teams heard about the situation, he began brainstorming with the workers. They reconfigured the bins and the job order so that needed parts would be closer. Then they measured the improvement to be between one and two seconds for each SUV passing through the station. "One improvement in one or two seconds—there's not a lot you can do with that," the supervisor said. "But two or three improvements can add up to significant savings."

Such obsessive attention to detail is at the heart of the productivity revolution that has swept corporate America. The goal in most cases is not to audit all employees' moves but to improve the processes that they know best. "Some people think little things like this are a bunch of nonsense," acknowledged one factory manager. "But I'm not looking for quantum leaps, I'm looking for incremental changes. Every little bit adds up." The payoff so far has been that, in six years of production, the M-Class SUVs made in Tuscaloosa have been among the automaker's most profitable vehicles.

Productivity gains are the fuel nudging the economy forward. After years of slow growth, the rate of output per hour in the United States began to accelerate in 1995, a dividend of the digital age. The past few years or so have been a productivity feast with annualized gains of nearly 4 percent, more than twice the historical average. The ability to produce more with less is what boosts incomes, increases corporate profits, and raises the overall standards of living.

In the auto industry, productivity gains of 6 percent to 7 percent a year—well above average—have helped lower the cost of a new car from about twenty-nine weeks of median pay in 1990, to about twenty weeks today. At the same time, auto industry wages, at about $20 an hour, have remained high for manufacturing. "If the productivity is there, we will keep the jobs here," said an economist, "otherwise, the jobs would have migrated to Bangladesh."

That sentiment reaches all the way to the factory floor in the Deep South, where there was virtually no auto industry until Mercedes arrived in 1994. "If we constantly do better, we get to keep our jobs," said a process engineer who joined Mercedes after graduating from the nearby University of Alabama. That mentality is widespread among workers. The gleaming Mercedes campus includes a health club and a day care center. And in 2006, based on the plant's performance, each of the factory's 2,000 employees received a bonus that could cover a year's tuition at a major university or a comfortable family vacation.

Mercedes started from scratch when it began building its Alabama factory in 1995. That allowed the company to incorporate modern "just-in-time" and "just-in-sequence" manufacturing techniques, setting up shop without the huge warehouses typically needed for storing parts. Instead of buying parts from thousands of suppliers and bringing them together at the factory, Mercedes had a hundred top suppliers preassemble major components such as interior consoles. Those suppliers are then tied into computers that track order flow.

Along the assembly line, plastic tubs arrived with mirrors, bumpers, or interior inserts at the point where they were needed, often within an hour of when they were plunked onto a chassis. The only storage is a series of shelves that look like a section of an Ikea store. The streamlined system helped save about $75 million a year. With suppliers handling much of the manufacturing, the Tuscaloosa plant is more like a final assembly plant.

The factory opened in 1997, and accepted 60,000 applications for 2,000 jobs. This allowed Mercedes to tap into the most qualified workers in the state. In its first full production year, output was about 68,000 vehicles. But Mercedes officials knew that they would have to increase output because of the popularity of SUVs. Production at the plant rose to more than 77,000 units in 1999, and then leveled off at about 80,000 vehicles in 2000 and 2001—all with the same number of workers. But another boost in output was needed. In the summer of 2002, a contract to produce several thousand M-Class vehicles a year at a factory in Austria was set to expire. That volume would shift to the Alabama plant. So early last year, the plant manager called for a *takt down,* a German term for reducing the time it takes each vehicle to move through assembly. Workers were required to come up with more efficient ways to do their jobs, with the goal of shaving six seconds off the time that vehicles spent at each of hundreds of workstations. Doing so was supposed to raise the number of vehicles produced each shift. Line workers and supervisors scrutinized slant walks, ergonomics, laborsaving gizmos, and safety enhancements that could reduce the number of accidents that shut down the line. Workers admitted that most of the improvements were small and barely noticeable on the line.

"The *takt down* was a phenomenal undertaking," recalled a line supervisor. "The first time, it tears people apart mentally. You do something different and you're going to get resistance. The team members got involved and made the difference." After nearly a month and more than 1,000 tweaks, the *takt down* became a success. By the end of 2002, the plant produced 88,271 vehicles, a 10.4 percent increase over the prior year. As a sign of the plant's success, Mercedes decided to build an adjacent sister plant to produce a new vehicle called the Grand Sports Tourer, which created another 2,000 jobs.

1. Assume the L-Class SUV plant has fifty managers out of a workforce of 2,000. Why can't these managers determine the best way to do things? Isn't that what they're paid for?

2. Isn't the L-Class SUV factory really a sweatshop? What if management comes back two years later and asks them to produce more with less? Is that fair?

3. What has productivity got to do with a company's profitability?

4. How many tweaks did it take for the *takt down* to be a success?

5. How can management encourage teamwork and employee suggestions?

6. An employee has been doing her assembly line job for seven years. She thinks the system is working just fine. What is so hard about thinking up new ways to do work?

7. Assume productivity does not increase at the plant for several years and the vehicles manufactured there are barely sold at a profit. List the possible options of the manufacturer for this manufacturing site.

CHAPTER 13

The Cost of Quality

Learning Objectives

After completing this chapter, you should be able to:

- *Explain why continuous improvement is needed to reduce or eliminate waste.*

- *List four categories of work.*

- *Describe three tools that can be used to eliminate waste.*

- *Explain what the term* hidden factory *means.*

- *List four costs of quality.*

- *Explain how quality should be measured.*

INTRODUCTION

Everything has an expense (cost). There is no such thing as an "expense-free" mistake. As an example, you print out a one-page document and notice you have misspelled one word. You return to the computer, make and save the corrected version, and then print it out. The action required less than two minutes of effort. The cost of the lack of quality was time, ink, and paper. Literally, this was nearly an expense-free mistake but the minor mistakes in a month by a company of 300 employees totaled could result in a sizable bill. We can exclaim that we are only human and are going to make little mistakes like this daily. True, but remember the quest for zero defects begins with *attitude*—an attitude that does not accept mistakes (defects) and seeks constantly to minimize them.

In this chapter, we will look at the cost of quality (COQ), which is (1) the cost of creating and maintaining a preventive quality system, and (2) the cost of poor quality (COPQ), usually represented by wasted raw materials, rework, and lost customers.

WORK

We all have jobs; in fact, we probably have had at least three or four different jobs in our lives. And we have all been involved in jobs (our work) that were very inefficient (with a lot of waste). Waste can come in many forms, such as:

- Wasted material
- Wasted capital (money)
- Wasted opportunities
- Wasted time and talent

Organizations waste human talent when they do not use the brains, time, and energy of all the people involved in a process. They waste opportunities when they lose a sale and the gross margin associated with that sale. Where does all this waste originate? It comes from work—what people work on and how they work. Waste does not involve just the work of people. It involves the work of the process equipment, machines, computers, electricity, chemicals, and so on. If people and machines are going to do work, it must be ***value-added work***. This means work that really adds value from the point of view of the external customer for the product or service.

Value-Added Work

Every business depends upon what people work on and how they work. As the Japanese discovered with Dr. Deming's help, organizations should work only those things that add value to external customers—the people who buy the products or services. Only if work increases the worth of a product or service for the customer is it considered value-added work. Examples of value-added work include adding a remote to a television, Internet access to a cell phone, or a new product to the company's line. Every person performing value-added work has a customer. If the work does not provide something an external customer needs or wants, it doesn't add value. Work that does not add value merely consumes time and resources. Thus, work that does not add value increases the price to external customers and reduces the efficiency of the producer.

All companies conduct some work (perhaps 5 percent to 15 percent) that is necessary, even if it does not directly add value, such as documenting sales and filing those documents. But companies strive to minimize any work that does not add value from the customers' viewpoints because customers only want to pay for work that adds value to what they buy. Imagine if you were required to keep records of all the work your organization does and you had to analyze those records to show how much is value-added work and how much is waste from the viewpoint of your outside customers. Then you submit those records along with your invoice to the customer. The customer may cut the invoice by 40 percent because the customer only wanted to pay for the value-added work, not the other work. Customers are willing to pay for necessary but not value-added work only because your competitors also charge for it. But some day that competitor may eliminate or greatly reduce any work that does not add value, become more competitive, and lower its prices. Companies must think about every task they complete from the external customer's point of view.

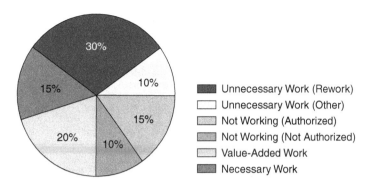

Figure 13.1   Categories of Work by Percent of Time

Unnecessary Work

Unnecessary work usually accounts for about 40 percent of all work, and about three-fourths of unnecessary work is rework—correcting or doing things over that weren't done right. Rework also includes the inspection necessary to find errors on work that was done wrong the first time. Because three-fourths of unnecessary work is rework, that means rework is 30 percent of all work. During about 25 percent of payroll hours in most organizations, people are not working. This includes legitimate nonworking time, such as vacations, holidays, and breaks. But it also includes a lot of time waiting or doing busywork due to scheduling problems or machine downtime.

Let's put all this together in a pie chart (see **Figure 13.1**). A typical organization expends 40 percent of its time on unnecessary work, 5 percent to 15 percent on necessary but not value-added work, and 25 percent of the payroll hours not working. This leaves less than 30 percent of the time spent on real value-added work. Only 20 percent to 30 percent of the time is spent on activities that add value from the customer's viewpoint. In the 1990s, some studies revealed that only 10 percent of the work done in the Western world is value-added work.

Eliminating Unnecessary Work

Experts on the work an organization does are the organization's customers, suppliers, and workers—the people involved in the work processes day by day. The only way to eliminate all this wasted work is by persuading these experts that improving the work processes is to their advantage. Organizations should reward people who look for change and are creative in eliminating waste. Too many companies reward those who supervise the greatest number of people. It is human nature for people to want to believe that the work they do has value. Workers know when their work is superfluous, and nothing is more damaging to self-esteem than to work continually at tasks that have no value to anyone. Ensuring that everyone is doing necessary and valuable work improves morale.

Organizations should study work to maximize time spent in work that adds value. Work is defined as a set of tasks performed by people, machines, energy, chemicals, and so on, to meet an objective. Work is measured in terms of the time it takes, what it costs, and the quality of the work product. To measure the time taken, the organization breaks down the work into tasks and subtasks. To analyze work, management might break down the six categories as shown in **Figure 13.2.**

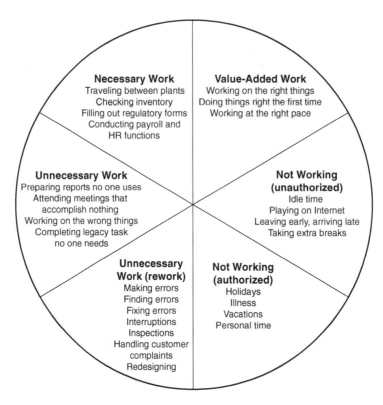

Figure 13.2 Breakdown of Categories of Work

Management should determine what work they pay people to do and how long it takes them to do that work. The object of the job analysis is for management, as well as the people doing the work, to understand what is being done, how it is being done, and the cost and value of those tasks to the company. The first step is to analyze in great detail all the work presently being done. Organizations must understand what tasks are being performed before they can eliminate waste. They must break down that work into bite-size pieces and examine work processes to see what can be eliminated, combined, reduced, or done in a better way. In realizing that the company is trying to eliminate waste, some employees may be reluctant to participate because a more efficient company may require fewer workers. On the other hand, a more efficient company may gain more market share and have to increase production and hire more workers.

Because much "necessary" work that does not add value is performed for internal customers, this work should be analyzed closely to determine whether the internal customer really needs this work. The definitions of the kinds of work—value-added, necessary but not value-added, and unnecessary—are also extremely important to how people think about work every day. Rather than think about work as a set of tasks being performed, people should continually examine those tasks so that they automatically classify them into the work categories mentioned. A company can begin the task of analyzing work by asking the following questions:

- Is the task necessary for the company?
- What is the task's value for the company?

- Is the work being done by the right part of the organization and the right people?
- Can the task be automated?
- Can the frequency of the service be lowered?
- Does eliminating the task pose a risk?
- Can the processes be improved?

Question constantly, keeping in mind that the questions have the objective of eliminating unnecessary tasks and the resulting waste.

THE WAR ON WASTE

One of the principal enemies of an organization seeking to maximize profits and gain market share is **waste**—anything that need not or should not have happened. *Waste* is just another way of saying *poor quality*. In a perfect world, there is no waste, but a perfect world is a fairy tale. In the real world, waste and errors are everywhere, especially in business, and waste and error are expensive (see **Figure 13.3**). The cost of poor quality (COPQ) is significant. Every day, people make errors and equipment malfunctions. The following are examples of waste:

- Wasted money on purchased products that aren't reliable
- Wasted creativity in not implementing good ideas
- Wasted time and energy on documents or meetings that don't get to the point
- Wasted organizational resources due to poor planning
- Wasted reputation by not meeting commitments

It would be wonderful if a silver bullet could win the war in eliminating the cost of poor quality. Unfortunately, no one approach will do the job. It takes a well-designed quality system and well-trained people.

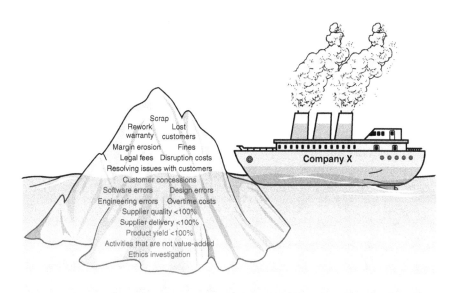

Figure 13.3　Iceberg Model of the Cost of Poor Quality

Measuring Waste

Many companies think that they have good quality systems if they merely have quality systems in place and if they address the known quality requirements. This is oversimplification, especially if quality for a process is defined as meeting the customers' requirements at the lowest price. Just having a working quality system in place is literally a bare minimum. Consider the fact that many companies (and their employees) do not have a handle on questions as basic as the following:

- How much raw materials/feed has to be scrapped during the manufacturing process because it does not meet requirements, and what is this costing?
- How much material has to be reworked or repaired (either during the manufacturing process or after delivery to the customer), and how much does this cost?
- What are the largest areas of scrap, rework, and repair?
- Which of these items should the company work on first?
- Is the company working on any of these items?
- Is the company getting better or worse at eliminating this waste?

The bottom line is that many companies simply don't have a handle on their quality and the cost of their quality. When asked how much of operations is being wasted due to scrap and rework, senior management of any American company (and probably any foreign company) will usually answer confidently, "Very little." Or they might answer, "Too much," meaning any waste is too much, but they still do not have an approximation of the dollar value of the waste occurring in their company. They don't know how much of their efforts are dedicated to making salable product and how much to reworking or repairing nonconforming product.

The Hidden Factory

During our development as an industrialized nation, American managers accepted the fact that inspection was the key to quality (see **Figure 13.4**). It was acceptable that one group of workers would manufacture the product, and then a second group of workers would inspect the output of the first group. The purpose of inspection is to evaluate the product's conformance to requirements. This concept served industrial America well for most of its existence, but is now considered unacceptable. Any organization that attempts to operate in this manner is doomed to failure unless it is a monopoly. Even then, such a company will

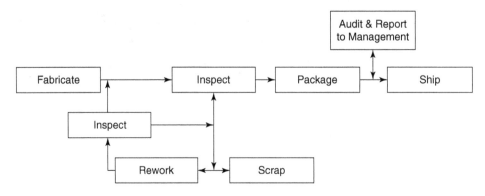

Figure 13.4 Manufacturing Inspection Process

never realize its full profit or quality potential. A company that operates like this has divorced the evaluation function from the production function, which allows waste. Let us look at the consequences of such a production system.

Companies that rely on inspection are in the business of separating good products from bad. These companies are detection-oriented (as opposed to prevention-oriented), seeking to detect nonconformances instead of preventing them. A nonconformance is any deviation from requirements. This includes machined parts that do not meet dimensional requirements, discrepant material purchased from suppliers, finished assemblies that do not pass acceptance tests, items returned from customers, and so on.

What do most companies do once they have detected products that do not meet requirements? Or, worse yet, what do they do when their customers detect these nonconformances? Typically, such companies rely on the efforts of a third group of people whose efforts create a unit in the company that is often referred to as the **hidden factory**. After the production workers build the product the first time and the inspectors evaluate the output of the first group and weed out defective products from good ones, a third group of workers are usually dedicated solely to rework. Sometimes the production group becomes the third group as they rerun off-specification material to bring it back on line. They spend a portion of their time reworking a product they had previously run through the process unit.

These are the people who constitute what used to be called, and sometimes is still called, the hidden factory. The people themselves are not hidden, and it is usually obvious they are reworking product. What is usually not so obvious is that these people constitute an operational workforce that is not listed on the organizational chart. Their rework and repair activities are usually not cost-isolated from the rest of the factory so that the work they do and the resources they use is not on any expense ledger. Their work, from both financial and productivity perspectives, becomes hidden.

For example, say a process unit goes off-specification for five hours. Everything the operators make during that five hours goes into an off-spec tank. The unit corrects the operational problem and production goes back on specification. Unit feed is lined up to the off-spec tank and the material in the off-spec tank is reprocessed through the unit, requiring five hours. **Table 13.1** reveals the expenses incurred.

Table 13.1   Off-Specification Expenses

1. Four operators at $32 per hour for five hours
2. One unit engineer at $55 per hour for five hours
3. Utilities (steam, electricity, cooling water, instrument and plant air, and cooling water) at approximately $1,425.00
4. Incident report (two workers for two hours at $32 per hour)
5. Total cost of incident $2468
Note: The new unit is now five hours behind in its production schedule. All debit costs are on a 24-hour-day cycle at the design rate.

The workers listed in Table 13.1 are, in most companies, a part of the hidden factory. They are paid out of the operational budget to do rework, not production. This example is a simple scenario. Imagine the cost if this unit was producing feed for a sister unit and the unit had to circulate instead of making product.

How large and costly is the hidden factory? It is not believed reliable statistics exist for much of American industry, simply because most companies do not know what their rework costs are. However, the U.S. Department of Defense Reliability Analysis Center found that poor-quality costs comprise 15 percent to 50 percent of all business costs. A study by *USA Today* found that the cost of poor quality comprised 20 percent of gross sales for manufacturing organizations and 30 percent of gross sales for service industries. When questioned on this subject, many chief executives guess their rework content to be below 5 percent, which does not come near the previous findings. Most companies are so accustomed to rework, they fail to recognize it.

What does all this mean?

- Quality is a measurable characteristic.
- Quality measurement should be based on the quantity and costs of nonconformances.
- Poor quality raises costs unnecessarily and increases the cost of scrap and rework.
- Quality can be most efficiently improved by measuring nonconformances in terms of quantity and cost, and systematically eliminating the major nonconformances.

THE COST OF QUALITY (COQ)

The *cost of quality* is a widely used term that is commonly misunderstood. The *cost of quality* isn't the price of creating a quality product or service. It's the cost of *not* creating a quality product or service. Every time work is redone, the cost of quality increases. The book *Principles of Quality Costs,* published by the American Society for Quality (ASQ), defines quality costs as "a measure of costs specifically associated with the achievement or nonachievement of product or service quality." Product or service quality may be defined as product or service requirements as shown in a contract. Obvious examples of the cost of quality include:

- The reworking of a manufactured item.
- The retesting of an assembly.
- The rebuilding of a tool.
- The correction of a bank statement.
- The reworking of a service, such as the reprocessing of a loan operation or the replacement of a food order in a restaurant.
- The cost of exceeding customer requirements.
- The cost of lost opportunities.

In short, any cost that would not have been expended if quality were perfect contributes to the cost of quality. Taken together, these costs can drain a company of 20 percent to 30 percent of its revenue (see **Figure 13.5**). Key areas of waste in a company include material, capital, and time, of which time is the biggest cost.

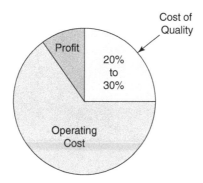

Figure 13.5   Cost of Quality as a Proportion of Revenue

THE MEASUREMENT OF QUALITY

Quality must be measured to manage it. Organizations must measure the results of how every part of the company is providing internal and external customers with their requirements. Whatever measure they choose should work for each department so that they can understand and assess the impact of individual decisions on the company as a whole and on customers. One way to do this is to measure the ***cost of quality*** (COQ). COQ is a dollar amount that represents the cost of avoiding, finding, making, or repairing defects in products or services. It is prevention costs, appraisal costs, and failure costs. The sum of these costs represents the difference between the actual cost of a product or service and what the reduced cost would be if there were no possibility of substandard service, failure of products, or defects in their manufacture.

We know we live in an imperfect world. It would be naive to assume any company could afford to spend nothing on quality and expect a quality product. However, if the goal is to provide a quality product, can one expect to make the costs of quality approach zero? Within practical limits, the answer is *yes*. When this happens, companies become more profitable and consumers receive better goods and services at lower prices. The practical limits are driven by costs associated with preventive quality measures intended to curtail larger detection-based costs. This will be explained in a later paragraph on preventive costs.

How should organizations capture and measure the cost of quality? Many of these concepts and statistics will be covered in later chapters when quality tools are introduced. In addition to these quality cost indexes, many organizations capture, segregate, and track quality costs in three areas: failure costs, appraisal costs, and preventive costs. The following is an explanation of each category.

Failure costs, which are often grouped under the acronym *cost of poor quality (COPQ)*, are those associated with the correction of nonconforming material, including scrap, rework, repair, warranty actions, and other costs related to the correction of nonconformances. Many organizations further subdivide this category into internal and external failure costs. One reason for doing so is that **internal failure costs** are a measure of a company's operating efficiencies, whereas **external failure costs** provide measures of both product quality and customer satisfaction. Ideally, failure costs should approach zero. In practice, failure costs typically comprise between 70 percent and 85 percent of an organization's total quality costs.

Internal failure costs include detecting errors after a product is made but before the product is shipped or a service is delivered. These are costs that take place inside the factory. A significant amount of internal failures indicates poor manufacturing efficiency, which may be due to poor training, poorly maintained equipment, inconsistent feed, and so on. Examples of internal failure costs include costs for the following:

- Scrap
- Rework
- Reinspection
- Retesting
- Material review
- Downgrading

External failure costs are costs that occur after delivery or shipment of the product—and during or after performance of a service—to the customer. These are usually much more expensive costs. (Remember the 1-10-100 Rule in Chapter 5?) The following are examples of these costs:

- Processing customer complaints
- Customer returns
- Warranty claims
- Product recalls

Appraisal costs are those related to the detection of defects. These are the costs associated with measuring, evaluating, or auditing products or services to assure conformance to quality standards and performance requirements. This cost category includes any measure used to separate good products from bad. Failure analysis and other activities focused on identifying underlying nonconformance causes should also be included in this category. Appraisal costs average around 15 percent of an organization's quality budget.

Appraisals are usually conducted after the product is produced but before it is released to the next recipient or shipped to the customer. Some appraisal costs could also fall into the category of prevention costs. The following are examples of appraisal costs:

- Incoming testing of purchased material
- In-process and final inspections or tests
- Product, process, or service audits
- Calibration of measuring and test equipment
- Associated supplies and materials

Preventive costs are the costs of all activities specifically designed to prevent defects or poor quality in products or services. This is the area that should dominate an organization's quality budget. Such costs include participation in the design process to eliminate potential failure modes, process improvements designed to prevent production of nonconforming product, process analyzers, statistical process control charts, training, and so on. Unfortunately, most organizations' preventive costs are relatively insignificant when compared to the failure and appraisal costs. The intent of any organization should be to lower failure and appraisal costs as a result of an intelligent investment in prevention-oriented activities. The following are examples of these costs:

- New product review
- Quality planning
- Supplier capability surveys
- Process capability evaluations
- Quality improvement team meetings
- Quality improvement projects
- Education and training

Presentation of the Cost of Quality

A picture is worth a thousand words. Most people resist reading a three- or four-page technical document. However, they will take the time to look at a one-page chart or graph. It is recommended that a company present its cost of quality data on a monthly basis in several formats. A pie chart is a simple but effective format. Another format, the Pareto chart, guides management in where to make improvements that will have big impacts, such as in rework. These indicators are quite helpful, as they allow management to visualize the comparative size of the hidden factory mentioned earlier in this chapter.

Assessing COQ can be an effective management tool because it helps a company compare the prevention, appraisal, and failure costs of one business process to another. It is an indicator of the effectiveness of the business. The higher the COQ is, the less effective the business is. Most American companies spend about 20 percent to 30 percent of their operating costs on the cost of quality. Steadily, they are bringing this number down. Through wise, persistent efforts, some companies have brought this figure below 10 percent. According to *Quality Digest,* a magazine for quality professionals, the overall quality of goods and services of many major American companies is as good or better than that of foreign competitors.

SUMMARY

In this chapter, you learned about the cost of quality (COQ), which is (1) the cost of creating and maintaining a preventive quality system, and (2) the cost of poor quality (COPQ), usually represented by wasted raw materials, rework, and lost customers. Every business depends on what people work on and how they work. Organizations should continue only those activities that add value to the external customer—the person who buys the product or service.

Many companies don't have a handle on their quality and the cost of their quality. They don't know how much of their efforts are dedicated to making salable product and how much to reworking or repairing nonconforming product. One of the principal enemies of any organization seeking to maximize profits and gain market share is waste. *Waste* is anything that need not or should not have happened. In many companies, waste and errors are everywhere, and that waste and error is expensive. People make errors, equipment malfunctions, and appliances break down the day after the warranty expires. Waste must be removed from processes through continuous improvement using quality tools.

REVIEW QUESTIONS

1. List four types of waste.

2. List four costs of quality.

3. State how much company revenue is consumed by the cost of quality.

4. Define the cost of quality.

5. List examples of the cost of quality.

6. Define *waste* and *value-added work*.

7. Explain why continuous improvement is needed to reduce or eliminate waste.

8. List four categories of work.

9. Describe three tools that can be used to eliminate waste.

10. Explain what the term *hidden factory* means.

11. Explain how quality should be measured.

12. List four examples of preventive quality costs.

13. List four examples of appraisal quality costs.

14. List four examples of failure quality costs.

15. Define *internal failure* and *external failure*.

16. Why should cost of quality be presented in charts or graphs?

GROUP ACTIVITIES

1. Read the following article, and then list seven losses and/or expenses incurred by Dell.

Dell Recalls Batteries Because of Fire Threat

By Mike Speegle (2008)

Dell said it recalled 4.1 million notebook computer batteries because they could erupt in flames. This was the largest safety recall in the history of the consumer electronics industry, the Consumer Product Safety Commission said. The recall raises broader questions about lithium-ion batteries, which are used in a host of devices like cell phones, portable power tools, camcorders, digital cameras, and MP3 players. The potential for such batteries to catch fire has been acknowledged for years and has prompted limited recalls in the past. But a number of recent fires involving notebook computers, some aboard planes, have brought renewed scrutiny. Dell reported to the safety agency that it documented six instances since December 2006, in which notebooks overheated or caught fire. None of the incidents caused injuries or death. Dell said the problems were a result of a manufacturing defect in batteries made by Sony.

The safety agency said the batteries were not unique to Dell, and that other companies using Sony batteries may also have to issue recalls. Sony has sold its batteries to most of the major computer makers. Depending on how many of the batteries are still in use, the cost of the recall could exceed $300 million. Dell refused to estimate the cost, but said the recall would not materially affect its profits. Sony, which affirmed that its batteries were responsible, said it was "financially supporting" Dell in the recall. The largest previous safety recall of a consumer electronics product, in October 2004, involved one million Kyocera cell phone batteries.

Dell has received several reports of burning laptops in recent months. In June, a Dell notebook burst into flames during a conference in a hotel in Osaka, Japan. In July, firefighters in Vernon Hills, Illinois, were called to the office of Tetra Pak, a food-processing and packaging company, to extinguish a notebook fire hot enough to burn the desk beneath it. That same month, a Dell notebook in the cab of a pickup in Nevada caught fire, igniting ammunition in the glove box and then the gas tanks, causing the truck to explode. Dell executives hoped the recall would prevent further damage to its image. Other computer makers that used Sony batteries took stock of their possible exposure to similar problems.

Lithium-ion batteries pack more energy in a smaller space than other types of batteries and are the cheapest form of battery chemistry. More powerful batteries are increasingly being used in more types of consumer products. This means that there is more likelihood for quality control problems and for design problems that could lead to future incidents and more recalls of these batteries. Federal regulations require that lithium-ion batteries be clearly marked with warnings when they are shipped in bulk on airplanes, and various agencies are considering more stringent regulations following a fire that was detected as a United Parcel Service cargo plane began its descent into an East Coast city. Though the cause of that fire, which consumed and destroyed the plane after it landed, has not been determined, lithium-ion batteries were suspected. No one was hurt. A single battery also caught fire in the overhead luggage bin of a Lufthansa passenger jet about to depart from O'Hare International Airport in Chicago. A flight engineer tossed it to the tarmac, where the fire was extinguished. Neither of the incidents involved Dell computers or Sony batteries. The Federal Aviation Administration lists three other incidents involving smoking or flaming lithium-ion batteries on cargo and passenger planes since 2004.

One Dell computer that caught on fire and was quickly put out was recovered from the fire and underwent tests. The unit worked when it was plugged into the power cord, despite the fire, which told the investigators that the problem was not with any circuitry or microchips. An X-ray of the battery pack revealed that the fire was not caused by an overcharged battery because a safety device was still intact. Rather, Dell said the cause of the fire was a short circuit in one of the batteries. It was caused by microscopic metal particles that contaminated the electrolyte, a porous insulator. Dell thought that the particles were released when the case of the cell was crimped near the end of Sony's manufacturing process. The same problem was associated with 22,000 notebooks Dell recalled in December.

Sony technicians provided additional data on all its batteries, not just those sold to Dell. They said data suggested a broader problem in the manufacturing process. Data did not show a predictable pattern for the batteries, which is why it was determined to remove them from the marketplace.

2. Read about and discuss the cost of quality described in **Table 13.2.** (Remember the 1-10-100 Rule?) This company makes schedule 80 PVC blanks that are cut into flange-like parts and used in the assembly of a final product that is sold to customers. After reading, complete the following:

    a. Calculate the original cost of the defect while it remains in the operator's area.
    b. Calculate the total cost of the defect once the customer has purchased it.

Table 13.2 Example of the Cost of Quality

Step	Description	Cost
1	The extrusion department makes 100 pounds of defective F-80 extruded flange blanks that are not detected. One area of the blank is bent slightly inward. Cost at this point is $1.25/pound.	$125.00
2	The blanks move through additional processes at a cost of $0.25/pound.	$25.00
3	An operator sets up a punch press to notch the ends. The setup requires 0.5 hours at $22.00/hour.	$11.00
4	An operator attempts to notch the blanks but is prevented by the bent area. The operator calls for a supervisor and waits 15 minutes. Cost is $22.00/hour.	$5.50
5	The operator and supervisor attempt unsuccessfully to process the blanks. A diemaker is called while the operator and supervisor wait. Total time is 20 minutes. The supervisor's time costs $30.00/hour.	$17.26
6	The supervisor, operator, and diemaker discuss the problem and attempt to process without success. They decide to grind out the die so the blanks will fill into it. The die is removed and taken to the die shop. Total time required is 1.0 hour. The diemaker's cost is $23.50/hour.	$75.50
7	The diemaker grinds out the die and sets up the press. Total time 3.4 hours.	$79.90
8	The diemaker, operator, and supervisor determine the status of the die, which seems to work. The diemaker and supervisor leave. The operator prepares to process blanks. Total time required 17.0 minutes.	$6.23

(Continued)

Table 13.2 *Continued*

Step	Description	Cost
9	The scheduler spends 1.5 hours at $15.00/hour to change the schedule and advise customers of the change.	$22.50
10	The operator processes blanks and carries them to the assembly department. Total time for the lot of 100-pound parts is 50 minutes.	$18.33
11	Workers in assembly work with the defective parts and expend 4.2 extra man-hours attempting to make the parts operate. The assembler time costs $11.50/hour.	$48.30
12	An inspector notices that some final products containing the blanks are difficult to operate. After twenty minutes, she summons the inspection supervisor. The inspector rate is $13.50/hour.	$4.45
13	The inspector supervisor, who makes $18.24/hour, arrives and the assembly supervisor, who makes $26.60/hour, comes to see what the problem is. The three meet on this problem for 13.0 minutes and decide to summon the engineer. All wait 11.0 minutes until the engineer arrives.	$22.22
14	The engineer, inspection supervisor, assembly supervisor, assembler, and inspector confer for 30.0 minutes and determine a way to fix the problem. All then depart. The engineer's time costs $36.00/hour.	$43.31
15	The inspector documents events, which takes 30 minutes.	$6.75
16	The assembler works for 3.9 hours to repair the finished products.	$44.85
17	After two weeks, a customer complains that some of the products caused problems shortly after installation. The customer requests a field service agent to fix the problems. After many phone calls, it is decided to send a field service agent to the construction site. Time involved in making this decision: Engineer 3.75 hours at $26.55/hour QA supervisor 1.75 hours at $19.25/hour Plant manager 1.45 hours at $60.00/hour Field engineer 2.00 hours at $42.00/hour	$304.25
18	A field service agent spends 36.0 hours in traveling to the site, fixing problems, and reporting. In addition, an engineer spends 1.6 hours on the problem. The sales manager spends 3.4 hours at $48.00/hour. Travel costs $1,115.00.	$1,345.00
20	The scrap value of the original 100 pounds of extruded blanks at $0.21/pound	$21.00
21	Cost of original error Final cost of error	$xxx.xx $xxxx.xx

CHAPTER 14

Quality Tools (Part 1)

Learning Objectives

After completing this chapter, you should be able to:

- *Explain the function of quality tools.*

- *Explain the use of the brainstorming tool.*

- *Explain the use of the check sheet.*

- *Construct a check sheet.*

- *Explain the use of the run chart.*

- *Construct a run chart.*

- *Explain the use of the scatter diagram.*

- *Explain the use of the flowchart.*

- *Construct a scatter diagram and flowchart.*

INTRODUCTION

No one works at a workplace where all the systems and subsystems are perfect. Many people have worked at sites where the same problems (defects, failures, and so on) have occurred year after year. They have learned to live with the problems. At first they probably responded typically by griping and complaining about the problems to management. But griping and complaining are not ways to effect change, as they merely irritate management.

175

Quality improvement and waste reduction can only be successful if the problems are measured (data) and then eliminated with quality tools. Remember management by data and facts—one of the four major elements of TQM—in Chapter 5 on total quality management? This chapter introduces you to five simple quality tools for collecting data and then assembling the data in such a way as to indicate facts. Quality tools stand on their own because a page of data with indicated facts laid in front of management is just facts; there are no personalities involved.

BEGINNING QUALITY IMPROVEMENT

Improving quality can be problematic because quality of many products and most services is a subjective attribute, which makes it difficult to identify and explain to others the major factors. Subjective attribute means that the determination is more of a judgment based on opinion than on factual data. Plus, recipients of feedback often view negative findings as personal attacks rather than sincere efforts to solve problems. When confronted with a quality problem, people typically react by defending themselves against personal attacks. The problem in need of improvement becomes secondary. At this point, good interpersonal skills and well-organized data become critical, or the problem will not be solved and may lead to new problems. Subjective attributes, such as opinions and attitudes, should not be used on quality problems.

One way to define the problem and to request the needed help is to think in terms of telling a story in a simple, objective manner so that others can understand the problem and take action to improve the situation. Statistical figures can be used to tell this story. No complex statistics or math formulae are needed; however, the statistics must be able to tell the story clearly and convincingly. The math and graph skills used in the majority of the quality tools and techniques in the next two chapters are not difficult to grasp.

Statistics

Statistics is a branch of mathematics that deals with the collection, analysis, and interpretation of masses of numerical data. In other words, a bunch of numbers and math formulae are used to determine what numbers reveal. An example of a simple statistic is the average height of fifteen boys. Statistics is data that are more objective than general statements.

For example, suppose a production manager discovers products were being shipped late from the site, and he stated this fact to the shipping manager. This could lead to an irritating or angry confrontation. However, instead of an aggressive confrontation with the shipping manager, suppose the production manager prepared a chart showing the number of late shipments per day. Along with the chart, a customer survey was included, in which key customers stated that the major problem in dealing with the firm was that shipments were late. Note that neither an individual nor team was singled out as the cause of the late shipments. It is hard to argue against facts because the data speak for themselves and remove personalities from the discussion. Statistical data display the problem in an objective, standardized fashion.

Some people think statistics is boring and hard to learn. However, people use statistics almost every day, even if they aren't aware of it, and would be startled to discover the importance of statistics. For instance, people make decisions based on statistics when they:

- Go on diets and weight-loss programs
- Determine bowling averages

- Make decisions based on weather forecasts
- Keep golf scores and handicaps
- Shop for groceries

Data and Data Collection

Before statistics can be assessed, data must be collected. Be careful when collecting data. Dr. Deming warned that collecting data that are easily available can become very expensive because it is ". . . simple . . . obvious . . . and wrong." The purpose of collecting data is to provide a basis for action. **Data** can be collected, analyzed, and then subjected to statistical methods to yield useful information for making decisions. Data gathering is a critical step, and can be a big source of waste in itself if not done thoughtfully and methodically. The following are three major reasons for collecting data about a problem:

- To reveal a problem exists
- To analyze the problem
- To minimize or remove the problem

The word *waste* could be substituted for *problem,* because the core activity in any business is to identify, quantify, and eliminate waste through process improvement. Historically, much of the data gathered about any process or system have come in the form of a final inspection in which inspectors asked questions such as: (1) Does a product meet specifications? (2) How many "good" ones are there, and how many "bad" ones? (3) Can the bad ones be reworked? Usually this type of inspection data reveals waste exists, but does little to help analyze or prevent it. This type of data reveals results, not causes. To assess the causes of waste, data on the process must be gathered. To do something about the defective process, the process that causes the defect must be analyzed. It is time-consuming work but, if you can't discover in detail what is causing the problems, it is unlikely you'll be able to fix them. That is why it is important to be critical about a data-gathering plan.

Data are important to all processes for the purposes of continuous improvement and preventative action. Think of data as iron ore. As long as the iron remains in the ore, it is of no value; it is just a piece of rock. However, something of monetary value is created when the ore is crushed and heated, and the iron is extracted and made into something. Think of data as that iron ore and statistics as the process of converting the ore into a product (value).

This is why samples are collected and analyzed (for data) on a routine basis in the process industry and why so many instruments collect data that is uploaded to a mainframe for eventual analysis. The historical data can then be used to describe a plant's processes and products. It can also be used to:

- Infer what might be happening to the plant processes or products.
- Predict what could be done to improve the product or processes.

A process description is a summary and report of collected observations and data. When we make inferences, we draw conclusions from presented or observed data. When we make a prediction about a process, we create an argument from past to future. In other words, we begin by looking at known data and end by making educated guesses that we hope are correct.

Sampling

A *sample* is a small portion of a population. Samples are collected to infer meaning to a larger population. For example, say we want to know the chemical composition of a 500,000-gallon product tank before loading a dozen rail cars from it tomorrow morning. Sending the whole 500,000-gallons to the quality control laboratory for analysis is impractical and impossible. So, we will sample the tank. For the sample to truly represent the whole population—a *representative sample*—it must be representative of the whole population (all 500,000 gallons). It must represent what is in that tank at that time; otherwise the sample data can be misleading. Process engineers and/or chemists establish the sampling scheme and sample size for each sample on a unit. The important point an operator must remember is to follow the sampling procedure to collect a representative sample. If the operator does not, there will be uncertainty in the measured value that can mislead the process unit workers into thinking there might be a problem. This uncertainty will be expressed in the sample's accuracy and/or precision. *Precision* is a measure of how closely individual measurements agree with each other; *accuracy* refers to how closely measurements come to the "true" value.

Proper sampling techniques are extremely important to process units. Bad samples yield misleading analytical results. It is not uncommon for a lot of time and effort to be spent looking for "ghosts" created by bad samples. This is a waste of time and resources.

THE SCIENTIFIC APPROACH

The core of quality improvement methods can be summed up in two words: *scientific approach*. It isn't complicated. A *scientific approach* is really just a systematic way for individuals and teams to learn about processes. It means agreeing to make decisions based on data rather than hunches, to look for root causes of problems rather than react to superficial symptoms, and to seek permanent solutions rather than quick fixes. Companies within the process industry use statistical thinking and statistics all the time.

Assume a process is having a problem that is causing rework, extra work, customer complaints, and off-product material for you and your teammates. You decide you want to eliminate the problem. You can go about this in two ways. The first is to ask everyone's opinion and react to the majority opinion. Do you have any data or statistics that converted the data into information? No, you are reacting to a subjective attribute—an opinion—in this case, the majority opinion. The odds are that this problem is not going to be solved any time soon.

The scientific approach requires an objective, methodical approach. The group will be trained on and use the following procedure, or one very much like it:

1. State the problem and the reason for the process improvement.
   - State the effect of the problem, not the cause (avoid finger-pointing).
   - Ensure the problem is measurable.
   - Define the gap between "what is" and "what should be."
   - Good quality tools to use at this stage are trend charts, flowcharts, and control charts.
2. Describe the current situation.
   - This stage breaks down the problem into its component parts (equipment, training, supplies, and so on).
   - Good quality tools to use during this stage are histograms, Pareto charts, and run charts.

3. Analyze and identify the root cause(s) through the use of quality tools.
   - Seek the root cause(s) of the problem. Drill down past a "quick fix."
   - Good quality tools for this stage are the fishbone diagram, flowchart, process model, and scatter diagram.
4. Create an action plan with duties and dates.
   - Develop a detailed outline of how the problem will be corrected. The plan should answer who, what, when, and where. This should be a neat, well-organized document that all can understand and follow.
   - Communicate with all key players and departments.
5. Analyze the results to confirm that the problem has lessened or been eliminated.
   - The objective of this stage is to confirm the problem has been decreased or eliminated.
   - Results should be displayed in graphs or charts (run chart, histogram, for example).

A scientific approach may, but does not always, involve using sophisticated statistics and formulae.

THE PURPOSE OF QUALITY TOOLS

A quality tool can be a chart, graph, statistics, or a method of organizing or looking at things. Quality tools help individuals and teams to continuously improve their work processes (see **Figure 14.1**) by making improvements based on data and facts. There are many quality tools—more than fifty—but the five tools in this chapter and the five in the next are some of the most commonly used by process industries. They are simple to use and reveal important information without requiring much time or effort from the users. Each quality tool looks at data in a different way and usually reveals different information. In other words, a run chart will not reveal the same information as a flowchart or check sheet.

This chapter will address the following quality tools:

1. Brainstorming
2. Check sheet
3. Run chart
4. Scatter diagram
5. Flowchart

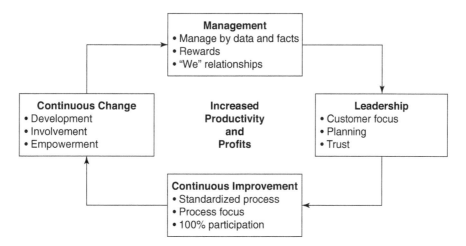

Figure 14.1   Continuous Improvement Cycle

This chapter and the following chapter provide the resources and knowledge to aid process technicians in accomplishing the following goals:

- Recognizing that improving business processes is everyone's job
- Basing troubleshooting decisions on factual information
- Developing and tracking a scorecard of measures to monitor performance and guide improvement

BRAINSTORMING

Brainstorming is a quality tool that generates a large number of ideas in a minimum of time. Working within a time limit to supply an idea or solve a problem, group members suggest as many ideas as possible and build on one another's ideas. Brainstorming is useful when a group needs to quickly generate a large number of ideas and consider many alternatives, and can help identify problems, explore causes, and develop solutions.

The following are two types of brainstorming methods:

● Freewheeling—Participants call out ideas spontaneously. Use this method when the goal is to gather creative ideas very quickly.

● Round-robin—Participants take turns contributing to an idea. They may pass if they do not have an idea at that time. Use this when the goal is full group participation.

Both methods continue until participants have no more ideas to contribute or time runs out. After brainstorming, use a selection method to limit the ideas to the most important ones.

Guidelines for Brainstorming

- Focus on one goal or problem at a time.
- Encourage all group members to participate and build on one another's ideas.
- Accept and welcome all ideas because they may eventually have merit or lead to ideas that do.
- No participant may criticize or humiliate another participant's suggestion.
- Permit clarifying questions only.

The object is to generate a lot of ideas as rapidly as possible—quantity first. The group can review the quality of the ideas later. The group leader should never allow an individual's idea to be criticized because it will tend to intimidate other group members from contributing their ideas. Valuable ideas might never be mentioned because of intimidation. Write down every idea quickly so that the team's energy and enthusiasm isn't lost waiting for the transcription. Once all members have contributed their ideas, the group evaluates them and arranges them by priority. Assume eighty ideas are written on a whiteboard or flip chart. Seventy of those may be rejected by consensus; then the group prioritizes the remaining ten from one to ten based on effort and cost. Idea number one will be tested first.

Procedure for Brainstorming

1. Make sure all participants understand the brainstorming subject.
2. Select the brainstorming method to be used.
3. Choose a leader and set a time limit.

4. Have the leader record all ideas on a flip chart or board.
5. Do not spend time discussing the ideas at this time.
6. Categorize and evaluate ideas.
7. Prioritize the ideas according to their effort and cost versus their impact on the problem.

CHECK SHEETS

Check sheets are easy ways to collect information. Check sheets can be used to count how often an event happens, measure the length of an event, or record the cost of an activity. Check sheets can be used for gathering data on how many times a pump trips, a shipping company arrives with an erroneous bill of laden, or how many times a control valve malfunctions. For example, an organization might want to record the number of requests for information it receives in a week. A check sheet can give a quick picture of the following:

- Which days are busiest
- How the requests were received (phone, mail, or in person)

Use check sheets when gathering data to count specific problems and to establish a pattern among those problems. Check sheets don't have to be on a standardized form. Create your own and use whatever is easiest for collecting and organizing data. The most effective check sheets are designed by employees familiar with the process or activity being investigated. What is critical is that information is collected in the same way if more than one person is using the same check sheet.

Guidelines for Check Sheets

1. Keep the form short and simple. Design the check sheet as a matrix, using the horizontal axis to measure time and the vertical axis to list the information being recorded.
2. Make sure the labels on the form are specific and there is room to record the data.
3. Enter information as check marks, tick marks, or some other form of measurement that can be totaled easily at the end of the recording period.
4. Provide space for comments for clarification.

Procedure for Check Sheets

1. Identify the specific data to be collected.
2. Decide how long the data will be collected.
3. Determine how the results will be used and analyzed.
4. Design a form for collecting the data.
5. Ask someone not involved in the design of the form to test it, then revise any confusing parts of the form.
6. Record the data consistently.

Example 14.1

Students have been complaining to the college cafeteria manager about the quality of food offered in the cafeteria. The manager sincerely wanted to address the problem, created a

Table 14.1 Cafeteria Food Quality Checklist

Complaint	Breakfast	Lunch	Supper	Total
Food is often lukewarm	II	III	I	6
Food is tasteless	IIIII III	I	II	11
Too much junk food, not enough healthy food	I	IIII IIII III	IIII	18
Food tastes "canned"	IIII II	II	I	10
Not enough variety in plate lunches		IIII II	IIII IIII	17
Everything is fried and greasy	IIII IIII	IIII IIIII II	III	24
Total	27	38	21	86

check sheet, and randomly asked students what they found dissatisfying about the cafeteria's quality of food. **Table 14.1** shows the results.

The cafeteria manager gathered a lot of good information on his check sheet. The three biggest complaints—too much junk food and fried and greasy food, and not enough plate lunch variety—give him a good starting point for improving his process (making desirable meals that produce a profit) and reducing customer complaints. Keep in mind the cafeteria must produce a profit, or why have one? The college could shut it down and put in five or six vending machines. Because the students are already on campus, it would be a shame to lose the opportunity of their business by ignoring their concerns.

RUN CHARTS

The next quality tool is the run chart. Run charts are simple graphs that track a process or activity over time to identify trends or shifts or to see the results of a change in a process. The horizontal axis measures time, and the vertical axis represents the frequency of some event. Information about a process or activity is plotted as data points on the graph.

A run chart is easy to use and understand. In addition, it shows the interrelation of data points, compares the values of data points over time, and helps to identify the normal variation of a process or activity. When used as a problem-solving tool, a run chart aids in organizing and analyzing data and identifying the root cause of a problem.

Guidelines for Run Charts

1. Chart the data in the order collected.
2. Record data as numbers, averages, or percentages.
3. Focus on long-term changes.

Procedure for Run Charts

1. Determine the specific process or activity to be monitored.
2. Collect the data using a check sheet or other counting method.
3. Create a graph showing the number of occurrences on the vertical axis and the units of time on the horizontal axis.

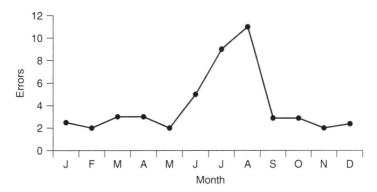

Figure 14.2   Run Chart of Shipping Errors by Month

    4. Plot the data points, and connect them with a solid line.
    5. Analyze the plotted information for changes and trends.

Example 14.2

To improve customer service and minimize complaints, the shipping department of a specialty chemical company located on the Texas Gulf Coast is trying to reduce loading and shipping errors of railcars, tank trucks, tote bins, and drums. Loading/shipping errors result in customer complaints, lost customers, and Department of Transportation (DOT) fines. Loading/shipping errors consist of the following: wrong DOT labels, wrong packaging, wrong product, improper container, wrong quantity, and late shipment. To get a picture of past performance and spot trends, a department quality improvement team created the run chart in **Figure 14.2** to analyze shipping errors over the past one-year period. All shipping errors for the month were totaled.

The data reveal a high occurrence of errors beginning in mid-June through August, which everyone knew were the hottest days of the year. Discussion with the loaders revealed they had little protection from the sun on the loading rack while loading rail cars and tank trucks. The drumming shed where drums, tote bins, and barrels were filled was a metal building open on both ends. It had a small overhead fan about four feet above the drumming station. Summer days easily had temperatures in the mid-90s, the humidity was high, and wearing personal protective equipment (PPE) while loading or drumming made the operators miserably hot. They just wanted to get the job done and go inside and cool off. Plus, mosquitoes were serious nuisances as the plant is located next to a salt marsh. Even if the operators wore insect repellant, the mosquitoes hung in front of the eyes in clouds. They had difficulty paying attention to detail between the high heat, humidity, and swarms of mosquitoes.

When this information became available, management extended the roof of the loading rack, protecting loaders from the sun, plus installed fans on the rack. In the drumming shed, the ceiling fan was lowered. Finally, management contracted a pesticide company to fog the plant for mosquitoes just before daybreak and two hours after dark.

Example 14.3

The safety department of a large refinery was concerned about the increase in workplace injuries. It decided to track the number of injuries on a run chart to get a picture of injury

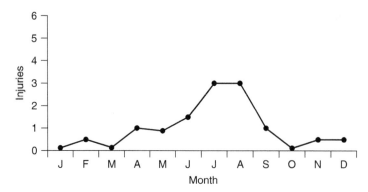

Figure 14.3   Run Chart of Workplace Injuries by Month

frequency by month. The safety department created the run chart displayed in **Figure 14.3** from data in the following table:

J	F	M	Apr	M	J	Jul	A	S	O	N	D
0.1	0.5	0.0	1.0	0.7	1.5	3.0	3.0	1.0	0.0	0.5	0.5

Management discovered that the heat and mosquitoes in the hot summer months also caused the higher level of personal injuries. Operators became frustrated, tired, and less cautious, resulting in injuries. Note: A lot of information could be obtained from the data or from mining just a little more data. For instance, the months could be grouped into seasons to determine if a spike in injuries was due to seasonal conditions such as hot or cold weather. They could also break the injuries into categories, such as injuries to hands, feet, backs, arms, eyes, and so on.

SCATTER DIAGRAMS

A scatter diagram is a quality tool that tests whether a relationship exists between two variables in a process, by showing what happens to one variable when the other changes. For instance, if X and Y are the two variables being tested and we increase the temperature of X, a scatter diagram would help determine if Y increases or decreases in value. The results can indicate a possible cause-and-effect relationship. Data are plotted as points on a graph. The possible cause is displayed on the horizontal axis (X), and the possible effect on the vertical axis (Y). The resulting pattern helps determine the relationship between the two variables. Use the scatter diagram technique to identity relationships that affect each other.

A study of a scatter diagram's results may reveal one of several patterns:

● Points distributed more or less equally on the graph indicate no apparent relationship. In other words, if X increases and Y sometimes increases and sometimes decreases, there is no relationship. This called a no-correlation scatter diagram.

● If both variables increase at the same time, a cause-and-effect relationship might be present. Thus, if both X and Y increase, this confirms a theory that two variables are related. This is called a positive correlation.

● If one variable decreases while the other increases, the pattern is called a negative correlation.

Important information can be gained from the types of correlation, and the relationship should be investigated further to find an explanation for the relationship.

Guidelines for Scatter Diagrams

1. Recognize that scatter diagrams don't tell if one variable causes the other. They only tell if a relationship exists and how direct the relationship is.
2. Be sure to analyze patterns correctly. The more the data points form a straight line, the stronger the relationship between the two variables.
3. Collect as much data as possible for the clearest, most accurate results.
4. Test other variables or relationships if the scatter diagram reveals no correlation.

Procedure for Scatter Diagrams

1. Identity the two variables to be tested.
2. Gather the data.
3. Draw a graph with the "cause" variable on the horizontal axis and the "effect" variable on the vertical axis.
4. Plot the data points. If values repeat, circle those data points as many times as appropriate.
5. Analyze the results.

Example 14.4

The shipping supervisor of the small specialty chemical company decided to use a scatter diagram to determine whether a relationship existed between the number of loading/shipping errors and amount of railcars the loader/shippers received per shift. Was there an optimum number that, if exceeded, would cause more errors? He created a scatter diagram (see **Figure 14.4**) from the following data:

X	3	4	3	4	5	6	7	8	9	10
Y	0	0.5	1	1.5	2	2.4	2.7	3.3	4	5.2

A clearly established relationship existed between the number of railcars loaded per shift and the loading/shipping errors per shift. What could cause this correlation? Probably a

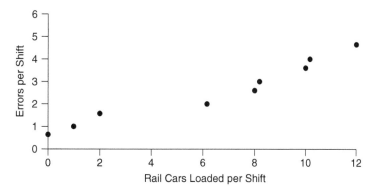

Figure 14.4   Positive Correlation: Loading Errors versus Tank Cars Per Shift

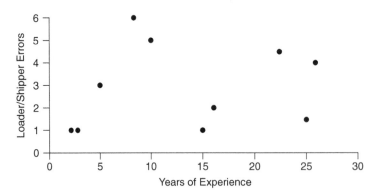

Figure 14.5 No Correlation: Errors versus Years of Experience

certain maximum number can be loaded per shift before a sense of "Hurry up and get the work done on time" begins to cause the operator to make errors. One possible way to avoid this is to add extra help when the count of cars to be loaded exceeds a certain number.

Example 14.5
Investigating further, the shipping supervisor decided to create a scatter diagram to determine if years of experience in loading/shipping had a relationship to the number of errors per loader/shipper. He did this hoping it might cast some light on the effectiveness of the current loader/shipper training process. He created a scatter diagram (see **Figure 14.5**) from the following data:

X	10	2	17	22	25	15	26	3	9	5
Y	5	1	2	4.5	1.5	1	4	1	6	3

The scatter diagram revealed no correlation of any type between the number of errors per shipper/loader and years of experience. This implied that experience could not compensate for the lack of knowledge and skills of a good loading and shipping training program. As has often been stated in the processing industry, "He's got seventeen years operating experience and it's seventeen years of doing the same things over and over again, including the wrong things."

Example 14.6
Investigating further, the shipping supervisor decided to create a scatter diagram to determine if the number of errors had a relationship to the hours of training each loader/shipper had received. He realized they had cut back on their loader/shipper training program years ago due to cost-cutting measures. Two of his best loaders had retired. New employees and transfers were being trained in the loading/shipping area by whoever was working the area. The training was on-the-job training. If that trainer had poor loading/shipping skills and knowledge, he would pass on that lack of knowledge and skills. The shipping supervisor created a scatter diagram (see **Figure 14.6**) from the following data:

X	36	32	28	24	20	16	12	8	4
Y	0	1	2	3	3.3	4	5	5	6

The shipping supervisor immediately saw the negative correlation between loading/shipping errors and training hours. Errors decreased as training hours increased. He realized he had

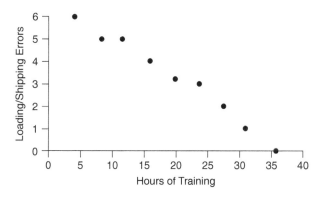

Figure 14.6 Negative Correlation: Errors versus Training Hours

to reinstate the loading/shipping training program. He now had statistics and data in hand to show his supervisor and to justify the added expenditure of a formal training program.

PROCESS FLOWCHARTS

Process flowcharts are quality tools that can be used to identify potential trouble spots, redundancies, and unnecessary complexities in almost any process. A process flowchart uses symbols to show the sequence of steps in a process or complex task. A process flowchart is like a map of the process that tracks a process from start to finish. In problem solving, a process flowchart quickly and simply describes a process so that all can understand it. Process flowcharts also aid planning by establishing a flow of events that can be used for resource scheduling.

Workers can use process flowcharts to analyze a process and/or establish workflow. By breaking down a process into logical steps, a team can pinpoint areas that need corrective action or identify potential problem areas for preventive action. A large process can result in a very complicated and confusing flowchart. Sometimes breaking a large process into three or four small processes is best. For example, the loading and shipping group might break a process flow into (1) the loading and DOT requirements of tank trucks, (2) the loading and DOT requirements of railcars, and (3) the loading and DOT requirements of tote bins and drums. Thus, three process flowcharts represent the flow of work of loading and shipping. Besides being a good troubleshooting tool, process flowcharts are excellent training tools because they give a quick and simple overview of a process.

Guidelines for Flowcharts

1. Construct process flowcharts from either left to right or top to bottom.
2. Ask the people most familiar with the process to help create the flowchart.
3. Verify that the process flowchart reflects the actual process that exists rather than the desired process.
5. Limit the scope of the process flowchart to the part of the process suspected of causing the problem.

Procedure for Flowcharts

1. Identify the process or task to be analyzed.
2. Name the process flow diagram after the process being diagrammed (for example, Flow Diagram of Styrene Reactor Section).

187

3. Determine the starting and ending points.
4. Fill in the necessary steps between the starting and ending points.
5. Draw arrows to connect each step with the next until the process is completely diagrammed.
6. Analyze the process flowchart.
7. Develop a plan of action to correct problems or improve the workflow.

Symbols

Flowcharts contain several symbols, but we will illustrate only three. Each symbol has its own function, and one cannot be substituted for another.

A *rectangle* normally stands for a task. Inside a rectangle will be a very brief sentence that describes what happens at that point.

A *diamond* represents a decision point with a yes or no question. The decision (question) is written inside the diamond, and arrows lead away and are labeled *yes* or *no*.

An *oval* represents a beginning or endpoint in the flowchart. For instance, a flowchart for purchasing a hamburger at a McDonald's might begin with an oval with the following written inside it: "Customer drives up to speaker phone on outside menu."

These symbols are typical of most flowcharts; however, other symbols are sometimes used. **Figure 14.7** shows some of the symbols.

Example 14.7

Revisit that small specialty chemical company. The plant manager of the company is very concerned about the large number of customer complaints about product shipments. He asked the loading supervisor to study the loading/shipping process to determine the root cause(s) of customer complaints. The plant manager asked the loading supervisor to have a starting point for a loading/shipping improvement process on his desk within a week.

The loading supervisor met with the loading trainers of each crew, then with the loader/shippers. The information he found was startling. No one used the loading/shipping standard operating procedure (SOP) because it was outdated and had not been revised in eight years. Because of that, each trainer trained loaders slightly differently, but the difference was enough to be noticed when the loader worked overtime on another crew. In effect, the training process was not standardized, and each crew leader trained in any manner he felt and with whatever documents were available. The loading supervisor collated all this information and put together a flowchart of the loading/shipping training process as he felt it should be (see **Figure 14.8**). After having the loaders review it for comment, he began to build his loading/shipping training program based on the flow diagram.

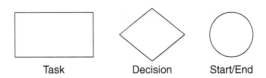

Figure 14.7   Flowchart Symbols

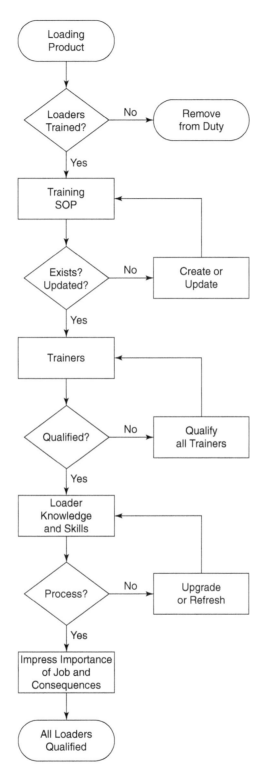

Figure 14.8   Loader/Shipper Training Process

SUMMARY

The problem with improving quality is that the quality of many products and most services is a subjective attribute (opinion), which makes identifying and explaining major factors to others difficult. Personality can play a role in receptivity, and recipients of negative findings often view them as personal attacks. But when objective data is placed in front of an individual, arguing against facts is hard because the data speaks for itself and removes personalities from the discussion. Statistical data displays the problem in an objective, standardized fashion.

The core of quality improvement methods can be summed up in two words: *scientific approach,* which is management by data and facts. Quality tools use this scientific approach to help individuals and teams continuously improve their work processes. A quality tool can be a chart, graph, statistics, or a method of organizing or looking at things. Companies have many quality tools they can use to develop statistics about their processes and make continuous improvements.

REVIEW QUESTIONS

1. Explain how data and statistics make changes easier.

2. Define a *representative sample.*

3. Explain what is meant by the "scientific approach."

4. Explain the function of quality tools.

5. Explain the difference between data and statistics.

6. List three purposes for data collection.

7. List the five quality tools in this chapter.

8. What would you use the brainstorming tool for?

9. The check sheet tool reveals _____.

10. The run chart tool reveals _____.

11. The scatter diagram tool reveals _____.

12. The flowchart tool reveals _____.

GROUP ACTIVITIES

Divide into small groups to address the following activities about the quality tools:

1. Brainstorm one of the following topics (or suggest an alternate to the instructor):

   - A better way to conduct road repairs without seriously affecting traffic flow
   - A weekly fitness schedule
   - A weight-loss plan

2. Create a scatter diagram between money spent on maintenance versus equipment breakdown. Plot money spent on the x-axis and equipment breakdown on the y-axis. What type of correlation is expressed?

Breakdowns	10	9	8	7	6	5	4	3	2	1
Thousands $	2.2	3.4	4.0	5.1	6.6	7.0	8.2	9.0	9.8	10.4

3. Create a check sheet from something in the classroom. Think up the subject of the check sheet and where to get the data. It might be something like the color of eyes of everyone in the room, race, gender, or ornamentation worn by everyone in the room.

4. Create a flow diagram of the following:

- Start at taking an order for a hamburger at the drive-in window and end at the customer driving off with the correct order.
- Consider a minor accident in which the driver bumped her head on the steering wheel, dented the front bumper and fender of her car, filed a claim, and eventually had the bodywork done on the car. The parties involved in the process are the other driver, witnesses, police department, automobile insurance agency, health insurance agency, claims adjuster, and body shop.

CHAPTER 15

Quality Tools (Part 2)

Learning Objectives

After completing this chapter, you should be able to:

- *Construct a pie chart.*

- *Explain the usefulness of the pie chart.*

- *Construct a cause-and-effect diagram.*

- *Explain the usefulness of the cause-and-effect diagram.*

- *Construct a histogram.*

- *Explain the usefulness of the histogram.*

- *Construct a Pareto chart.*

- *Explain the usefulness of the Pareto chart.*

- *Explain the usefulness of the control chart.*

INTRODUCTION

This chapter presents five more quality tools that will help individuals and teams to continuously improve their work processes. These tools are slightly more complex and require a little more time, but are still simple to use and reveal important information in different ways than the previously mentioned quality tools. This chapter will address the following quality tools:

1. Pie chart
2. Cause-and-effect diagram (fishbone diagram)
3. Histogram
4. Pareto chart
5. Control charts

Remember, the purpose of any quality tool is to convert data into useful information that will improve the process (system) and eliminate waste.

PIE CHARTS

Pie charts are circular graphs in which the sections of the circle represent the proportional amounts of relative quantities. In general, pie charts represent percentages with all the sections of the chart adding up to 100 percent. Use this tool to graphically represent percentage data. It is a powerful tool that reveals important values immediately and is widely used by the process industry.

Chart sections can be colored, with a color key added to explain the significance of the colors, or the information may be presented in each section of the chart. For example, see **Figure 15.1**, a pie chart of customer complaints made to company representatives.

Guidelines for Pie Charts

1. Ensure everyone involved in creating the pie charts agrees to the number of sections and the names of each section. Terms can mean different things to different people. For example, someone may want to identify or show examples of "Other" in Figure 15.1.

2. Keep the labels simple and specific to avoid misunderstandings.

3. Ensure team members are familiar with the process or action to be charted so they can contribute. However, having a team member who is unfamiliar doesn't hurt because this helps avoid a "silo" viewpoint. Silo viewpoints are narrow-minded viewpoints derived from fixed concepts of how things are or should be. Having a team member who is unfamiliar with the process allows them to think out of the box. Such people may see something from a different perspective.

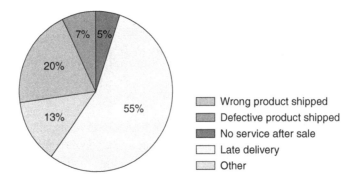

Figure 15.1 Pie Chart of Customer Complaints

Procedure for Pie Charts

1. Determine the data you are collecting.
2. Collect the data and sort them into categories (chart sections).
3. Convert the data into percentages.
4. Calculate the width of each section based on the percentage value of the section.
5. If using color, create a key in the corner of the graph to explain the meaning of the color.

CAUSE-AND-EFFECT DIAGRAMS (FISH BONE DIAGRAMS)

The cause-and-effect diagram, also called the **fishbone diagram**, was created by the Japanese quality expert Ishikawa. **Cause-and-effect diagrams** help people understand the complex relationship between an effect (a problem or goal) and its causes, and aids in problem solving. Often, cause-and-effect diagrams are used in conjunction with brainstorming. Teams can use brainstorming techniques to produce a list of the most probable causes after a problem (effect) is identified. The effect is written on the right side of the diagram and a horizontal line is drawn from it to the left side of the page (see **Figure 15.2**).

Categories of causes are put in boxes at the end of each rib that feeds into the horizontal line (backbone). Commonly selected categories are materials, machinery, measurement, methods, and people. You can use as many or as few categories as desired. Causes can fall into whatever categories the group determines. Other categories could be information, performance standards (quality, cost, and schedule), facilities, and training and knowledge. Each major category might have numerous causes and contributing factors (subcategories). Subcategories are then attached to each rib. The diagram's fishbone shape gives the cause-and-effect diagram the alternate name of "fishbone diagram."

Guidelines for Cause-and-Effect Diagrams

Use cause-and-effect diagrams to identify as many possible causes of problems—or goals—as possible and to identify areas from which to gather data.

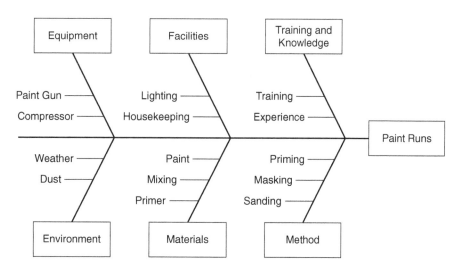

Figure 15.2   Cause-and-Effect Diagram of Paint Runs

1. Ensure group members agree on and understand the effect (problem).
2. Position the diagram so all group members can see and contribute to it.
3. Begin by looking at the "effect" and asking, "Why did this happen?" to identify major categories. Consider anything that might explain the effect.
4. Then ask, "Why?" again to identify the numerous subcategories under the major categories.
5. Don't evaluate ideas while generating possible causes.
6. Facilitate the discussion by asking specific *who, what, when, where, how,* or *why* questions.
7. Add to the diagram as additional information develops.
8. Use forms, such as check sheets, to collect information quickly.

Procedure for Cause-and-Effect Diagrams

1. Write a problem statement that clearly states the problem that exists or the goal to be attained. Be sure that everyone agrees on the problem statement. The problem statement could be something like, "Paint runs on cars."
2. Identify three to six major categories that might have led to this effect.
3. Brainstorm to fill in causes (subcategories) under each major category; connect each cause (subcategory) to the major category line (rib).
4. Brainstorm factors that could be contributing to each subcategory, and position these on a line moving away from the subcategory.
5. Discuss each factor and how it might contribute to the cause. List this information next to the factor.
6. Refine each category by asking the following questions:
   ● What causes this?
   ● Why does this condition exist?
7. Ask "why" numerous times.
8. Reach a consensus on and circle the most likely causes.
9. Look for causes that appear repeatedly.
10. Agree on steps to either collect data verifying causes or to eliminate causes through corrective action.

Example (See Figure 15.2)

The owner of an auto body shop had been in business for twenty-three years and rarely had customer complaints about his paint jobs. The owner was shocked to learn that the paint on two cars his shop had painted the previous week had paint runs. He used the cause-and-effect diagram in Figure 15.2 to identity possible causes, such as new paint or primer supplier, new worker, heavy rain (humidity) that day, and so on.

HISTOGRAMS

Histograms are important diagnostic tools because they give a clear picture of data distributions that might otherwise be difficult to visualize. A histogram is a very impressive way of suppressing irrelevant data while highlighting relevant data. A page or table full of numbers does not readily illustrate any distribution pattern. A histogram lets us visualize (see) a distribution pattern. However, if the data are grouped in certain increments, valuable information appears in the histogram. Histograms can help determine the need to correct a process or improve its efficiency. Histograms are best used to determine and display the data distribution for a process. This chapter is just introducing you to histograms, which will be studied and discussed in more detail in Chapter 17.

Guidelines for Histograms

1. Determine the optimum number of data group ranges to identify a pattern. Too few ranges might make a pattern difficult to detect, and too many will break it up.
2. Keep the ranges equally sized (bars equal in width). For instance, the ranges for temperature could be every two degrees Fahrenheit, and a range for flow could be every five gallons per minute.
3. Look for normal or unusual or unexpected distribution patterns.
4. An odd pattern can mean that a special cause is influencing the process or that there is a flaw in the measurement method.

Procedure for Histograms

1. Identify the measurement data to be charted (temperature, pressure, and so on).
2. Collect the data and count the number of times each measurement occurred, starting with the lowest measurement.
3. Label the horizontal axis with the measurement values.
4. List the frequency of each measurement on the vertical axis.
5. Draw vertical bars to the height of the corresponding frequency measurement for each value.
6. Analyze the chart by looking at the distribution. A histogram of a normal distribution (population) will look like a smooth bell-shaped curve.

Example 15.3

Remember the specialty chemical company with all its loading/shipping errors? The shipping supervisor did not want to lose his year-end bonus because of all the customer complaints. He had reestablished a loading/shipping training program but it would be a while before everyone was trained. Meanwhile, the plant manager would notice that loading/shipping errors were still occurring and customer complaints still rolling in. The shipping supervisor decided to make a list of all loading/shipping errors, and then use the list to create a histogram (see **Figure 15.3**). He grouped his data into container errors.

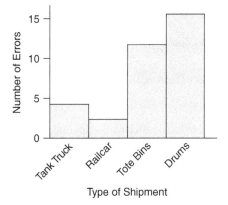

Figure 15.3   Histogram of Loading/Shipping Errors

The first thing he noticed was that the greatest amount of errors occurred with drums. He wondered how much customer complaints would be reduced if he created a short course that trained everyone on just loading, labeling, and filling out the bills of lading for drums. He got that answer with the next quality tool.

PARETO CHARTS

When focusing on the most serious problems or causes of a problem, use Pareto charts to rank information from most serious to least serious. A Pareto chart is a quality tool that lists information in terms of cost, percentage, or frequency, and displays it in a vertical bar graph format similar to a histogram. Use a Pareto chart to identify the causes of problems, to rank the relative importance of the problems, to determine a starting point in problem solving, or to assess improvement in a situation.

The principle behind the Pareto chart is called the 80/20 Rule, which states that 80 percent of the problems come from 20 percent of the causes. This means that addressing the problem or cause represented by the tallest bar achieves better results than working on problems or causes represented by shorter bars.

Guidelines for Pareto Charts

1. Draw each bar to scale on the Pareto chart.
2. Label the left vertical axis with the unit of measurement. Record raw data on the left vertical axis.
3. Show percentages on the right vertical axis.
4. Verify both axes are drawn to scale.
5. Draw a line across the tops of the bars showing the cumulative totals of the categories.
6. Label all parts of the chart clearly.

Procedure for Pareto Charts

1. Identity the causes or problems to be ranked.
2. Select the unit of measurement, such as cost, frequency, or percentage.
3. Determine the time period for recording the measurements.
4. Collect the data.
5. Rank the results from left to right on the horizontal axis, starting with the tallest bar.
6. Interpret the results. The items with the highest priority appear on the left.
7. Develop a plan of action.

Example 15.4

The shipping supervisor divided his customer complaints into the following categories:

● The customer received improperly labeled shipments.
● The customer received improperly packaged shipments.
● The customer received off-specification shipments.
● The customer's shipment was late.

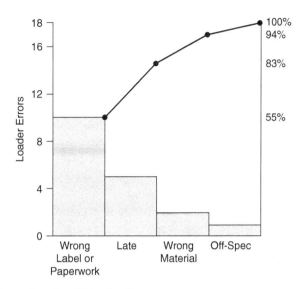

Figure 15.4   Drum Loading/Shipping Errors

The shipping supervisor tallied the drum complaints over a three-month period and found a total of eighteen complaints. The tally was as follows:

Wrong label or paperwork	10
Late shipment	5
Wrong material	2
Off-specification product	1

The histogram (see Figure 15.3) pointed out that the majority of shipping errors (and customer complaints) were due to drum shipments. Tote bin shipments followed as the next largest category of complaints. A quick estimate revealed that the supervisor could reduce his customer complaints by more than 80 percent if he could reduce the errors with those two types of shipments, which should guarantee his year-end bonus. First, he decided to concentrate on drum shipment errors, because they occurred most frequently. To better emphasize the data to the loaders, he constructed a Pareto chart to illustrate where the greatest problems were and where they should concentrate their efforts for improvement. The Pareto chart is shown in **Figure 15.4.** It shows that 55 percent of drum complaints could be eliminated if the loaders ensured the labeling and paperwork were correct. Then, if they solved the problem of sending the railcars to the buyer on time, they would have eliminated 83 percent of customer complaints about drums.

The supervisor and crew decided that training sessions were needed on (a) Department of Transportation (DOT) drum labeling and (b) correctly filling out the shipping forms. Meanwhile, the supervisor, thinking of that bonus check, started collecting data on the types of loading/shipping errors with tote bins. He wanted to eliminate that source of waste next.

CONTROL CHARTS

A critical quality tool for any manufacturing process is the statistical process control chart. The performance and stability of a process can be monitored and controlled by using control charts. Control charts help distinguish between **common** and **special causes** of variation in a process (variation will be discussed in detail in Chapter 16). In a stable process, random fluctuations result from common causes. Reacting to these normal variations will only make things worse. Because they are normal, they are part of the process and should be accepted. Nonrandom fluctuations indicate that a *special cause* (also called *assignable cause*) has changed the process. Special causes are the major sources of trouble in the production of product or delivery of a service, and this variation makes the output pattern fluctuate in an unnatural manner. Examples of special causes include failing or failed pumps, temperature controllers, pressure controllers, and so on. In other words, equipment is not working the way it is supposed to and has caused special or assignable cause to suddenly appear.

Control charts reveal when an abnormal change in a process has occurred due to a special cause, and they help to identify and correct the change. Control charts do not reveal what has changed or why it changed. With knowledge of the processes and equipment, team members determine this. Control charts will be addressed in more detail in Chapter 18.

A brief look at control charts reveals that all control charts have three lines:

● A centerline that provides the average value (temperature, pressure, flow, and so on) of the process.

● An upper line called the *upper control limit* (UCL), drawn at a calculated distance above the centerline. Any point above the upper control limit indicates that the process is out of control and in danger of making off-specification product. Special cause has entered the process.

● A lower line called the *lower control limit* (LCL), and drawn at a calculate distance below the centerline. Any point below the lower control limit indicates the process is out of control and in danger of making off-specification product. Special cause has entered the process.

There are different types of control charts. Two of the most common are charts for continuous variables and charts for discrete (attribute) data. Refining and petrochemical companies principally use charts for continuous variables. **Figure 15.5** shows a control chart for an organic compound from a chemical plant's wastewater stream that empties into a river. The plant has a permit that allows it to release a daily average of 70 parts per million (ppm).

The dotted lines are the upper and lower control limits (UCL and LCL). The solid centerline is the average. The chart reveals that the plant is releasing about 52 ppm on average. Its highest release in the last twenty-eight days was about 66 ppm; its lowest was about 41 ppm.

SPC helps to control the process by alerting (usually by an alarm) the operator of the intrusion of special cause. This allows the operator to take immediate corrective action the moment a value occurs outside the limits of the control chart. This is short-term corrective action. Over a longer period of time, the control chart gives information on the amount of fluctuations that can lead to long-term corrective action through new equipment or better procedures.

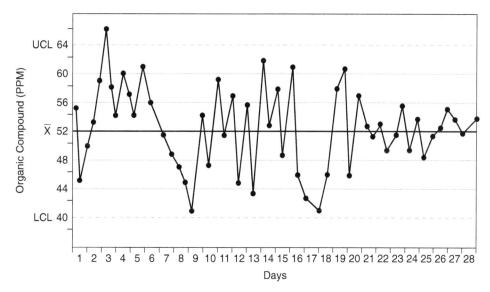

Figure 15.5 Control Chart Example

SUMMARY

This chapter introduced five more quality tools that can be used to continuously improve work processes and reduce waste.

The pie chart is a circular graph used to graphically represent percentage data. In general, pie charts represent percentages with all the sections of the chart adding up to 100 percent. The cause-and-effect diagram helps people understand the complex relationship between an effect (a problem or goal) and what can cause it. This diagram is used as a problem-solving tool. Histograms are important diagnostic tools because they give a clear picture of data distributions that might otherwise be difficult to visualize. Histograms are best used to determine and display the data distribution for a process. A Pareto chart is used when focusing on the most serious problems or causes of a problem. It ranks information. Control charts help distinguish between common and special causes of variation in a process. The performance and stability of a process can be monitored and controlled by using a control chart. Control charts reveal when an abnormal change in a process has occurred and help to identify and correct the change.

REVIEW QUESTIONS

1. Explain what a pie chart graphically represents.

2. The cause-and-effect diagram is used for _____.

3. The histogram is used for _____.

4. The Pareto chart is used for _____.

5. Explain the usefulness of the control chart.

Table 15.1   Traffic Fine Revenues for Three Months

January	February	March	April	May
45.00	35.00	120.00	75.00	25.50
50.00	49.50	75.00	69.50	75.50
18.00	18.00	110.50	25.00	20.00
15.00	25.00	70.00	19.50	30.00
22.50	49.50	55.50	15.00	55.00
49.00	26.00	100.00	33.00	52.50

GROUP ACTIVITIES

Divide the students into groups. Ask the groups to use each of the following quality tools:

1. Cause-and-effect diagram—Write an article for the campus newspaper about automobile safety. Address the problem of a driver losing control of his car. Use the cause-and-effect diagram to develop data for the article. The head of the diagram (effect) will be "Reasons for loss of control of car." Use as many categories as you wish, such as "weather" or "road conditions," and so on.

2. Histogram—A small town in Central Texas has a four-year college known as a "party" college. The sheriff's department did not receive a budget increase for the year, and the department needs some new equipment. The department gets 30 percent of its revenues from tickets. The sheriff, curious about how much revenue is generated by traffic tickets, decides to collect the data for the previous three months. The data are shown in **Table 15.1.** Construct a histogram from the data using the following distribution categories: parking tickets, $20 or less; failure to give a turn signal, $21 to $40; headlight or taillight not working, $41 to $65; running a stop sign or red light, $66 to $90; and speeding: greater than $90.

   What type of tickets do you think the sheriff will encourage his officers to write?

3. Read the following scenario:

Department Store Scenario

Jill Jones received a promotion to department manager of women's apparel at a major department store. The promotion was for a different store than where she was currently working. After a week on the job at the new location, she noticed the reoccurrence of things that reduced her department's efficiency, productivity, and profitability, not to mention reduced customer satisfaction. She decided to hold a meeting with her two assistant managers, Pam and Mary, and to try to understand the system (process) currently in place for her department.

The following is the meeting dialogue:

JILL: Pam, Mary, I called this meeting to get an understanding of how our department conducts its business. I want to understand the system we have in place to meet our objectives of sales, customer satisfaction, and continuous improvement.

Mary and Pam both nod and smile.

JILL: What I want you to do is explain the processes we use for (1) putting new styles on the racks and shelves immediately, (2) returning defective products for credit, (3) employee training, and (4) customer satisfaction.

PAM: I'll start off with defective merchandise. We instruct our clerks that when they find bad apparel, they are to pull it and put it on the vendor return table in the stock room.

JILL: And then what?

MARY: When we have time, we sort them by vendor, box them, and mail them back for credit. Sometimes we can get the vendor sales representative to take some back with them.

JILL: What is the average length of time an item might remain on the vendor return table?

MARY: Oh gosh, maybe three weeks.

JILL: So, there is no person responsible for that task and no timetable?

PAM: Right. That's the way it has always been done here.

JILL: What about new styles and merchandise? How do we get the merchandise to this store, to this department, and on the racks?

MARY: Merchandise is freighted via truck to the back dock and unloaded in the main storeroom.

PAM: When we have time and can remember—you see how hectic it is here—either Mary or I go to the storeroom and check for the arrival of new merchandise.

JILL: Isn't there a storeroom attendant or manager that could phone you when our department has arrivals? What about invoices notifying you of the arrival date?

MARY: The storeroom attendants sometimes notify us but they say that's not their job. Pam and I try not to make waves. As for invoices, things rarely arrive on the date specified on the invoice.

PAM: Now, if we have merchandise in the main storeroom we take a cart and bring it up in the freight elevator.

JILL: Who is *we*? And how promptly?

MARY: Me, Pam, and any of the clerks. We try to bring it up as fast as possible but it all depends on who is available at the time. To be honest, even if we got all the stuff up here on the day it arrived, we often don't have the manpower or time to get it on the shelves.

JILL (*taking notes furiously*): What type of training do we give our clerks?

MARY: Well, we have new hires do on-the-job training with a clerk. They learn to work the register, how to check inventory, how to handle customers, phone etiquette, the usual stuff.

JILL: I'd like to see copies of the training procedures and training records.

PAM (*giving Mary a startled look before looking back at Jill*): Well, we don't have training procedures or records. We just train people.

JILL: Who are the trainers?

PAM. Anyone with experience. Sometimes we do the training, sometimes a clerk does.

JILL: So, how do you know if a new hire is adequately trained?

MARY (*frowning*): Jill, it's just a clerk's position. There isn't much to it and anybody can learn it.

JILL (*nodding*): Who handles customer complaints?

PAM: Whomever the customer is complaining to. If it's a really tough customer, the clerks bring them to us.

JILL: Is everyone trained on customer satisfaction and how to handle customer complaints?

MARY: Well, I think so. I mean, we tell our people to always stay calm and smile, no matter how bad it gets.

JILL: Is there any follow-up on the customer's complaint to see if it was handled satisfactorily?

MARY (*frowning*): No.

JILL: What are our most common complaints and how frequent are they, say on a monthly basis?

PAM: Hmmm, I guess it would be items not in stock, or maybe the right size not in stock. Right, Mary?

MARY: Yes, I think that is the most common complaint.

JILL: You guess? Hmmm. Do we keep any statistics that tell us how we are doing besides the weekly and monthly sales report?

PAM: Statistics? No. What on earth would we need statistics for?

a. Create one or more flowcharts that represent the way the process should be. Because a large process can result in a very complicated and confusing flowchart, it is often best to break a large process into three or four small processes, such as employee training, customer complaints, and so on.

b. Draw a fishbone diagram that finds the root cause of the department's systemic problems. If you have more than one problem, use a fishbone diagram for each problem.

CHAPTER 16

Variation

Learning Objectives

After completing this chapter, you should be able to:

- *Explain why the worker is the expert on process variation.*

- *List the various causes of variation in the process industry.*

- *Explain how process variation can be used as a tool to improve a process.*

- *Explain the difference between* common cause *and* special cause *variation.*

- *Explain why carefully planning the data-gathering process is important.*

- *Describe Shewhart's two approaches to process improvement.*

- *Explain why the worker cannot do anything about common cause variation.*

- *Explain the importance of control charts for process improvement.*

INTRODUCTION

This chapter introduces the concept of process variation and its uses in the drive to eliminate waste through continuous improvement. **Variation** is the change or deviation, in form, condition, appearance or extent from a usual state, or from an assumed standard. Process variation—the deviation from a norm within a process—is a leading contributor to waste in work processes. It is also one of the principal tools that organizations can use to find and eliminate waste. Once people know how to understand or detect process variation, they can use it like a map to point the way to the parts of the process that need the most attention. That is why variation can guide workers to effect continuous improvements.

It is a fundamental fact that variation exists in nature and in every repetitive situation. Variation exists in raw materials, the processing operations, the human factor, inspection, testing and analysis, and product performance itself. The science of statistics generates a pattern of thinking and provides a system of strategies that offer the most effective method for understanding and controlling the variation that exists in all industrial processes.

VARIATION IN PROCESSES

As stated in the introduction of this chapter, variation is the change or deviation in form, condition, appearance, and extent from a usual state, or from an assumed standard. The less variation in a process, the more the worker has control of the process. Also, less waste results from less variation in a process. With less waste, efficiency improves, which in turn can translate into higher profits. Understanding, controlling, and reducing variation in all processes should be the goal of all employees.

Even with accurate measurement techniques, an organization can make little use of process variation unless employees look in the right places for the causes of the variation. Most managers assume that, once a process has been established, it will work the way it was designed to work unless the worker does something wrong. In other words, they assume that variation in a process is caused by and can be controlled by the people running the process. This thinking applies to all kinds of processes, whether manufacturing or services. For hundreds of years, people have believed this. If production yield drops from 92 percent to 88 percent, the people running the process—or the suppliers who supplied the parts—must be responsible.

Statistical evidence tells a different story. It says that, in a stable process, 90 percent to 95 percent of the process variation is inherent in the system and can't be controlled by the people working the system. Only 5 percent to 10 percent of the problems are due to people operating the system. Because management designs and controls the system, only management can address most of the system's problems. Workers cannot control some problems in the system.

Management's Role in Variation

Workers in a process system are in the best positions to identify problems of a flawed system, because they are working in the system daily and see problems when they occur. That is why Dr. Deming called the workers, "the experts." But they are not in positions to do much about the problems because they do not have the power to change the system. Either the managers must change the system, or the organization must empower the workers to do it.

Because many managers assumed that workers cause the problems, traditional management theories supposed that the worker must be tightly controlled and disciplined to improve performance. This reasoning resulted in the authoritative mind-set of the old system. The "old system" refers the authoritarian system that implied that managers were paid to think, and workers to work. In that system, workers were not required to think. The old system assumed that those in authority possess the important knowledge. The workers were considered to be simply imperfect tools required to operate the system. This was the old culture of "us versus them," or management versus labor. That culture began to be replaced in the 1980s by a culture of empowerment and mutual respect (see Table 5.3 in Chapter 5), which has some of its roots in the quality circles the Japanese established after the war.

Worker Experts

A new philosophy of work evolved from understanding that more than 90 percent of the problems in an organization can be fixed only by those working on the system, because they possess most of the knowledge of systemic problems. Improvements in the system come through cooperation between labor and management. The people working in the system are asked to identify the problems that cause waste and to assist managers in fixing the problems. Because both groups need to work together to make continuous improvements, management needs to develop a spirit of teamwork to achieve the cooperation of all concerned. Managers must listen to their worker "experts" and provide them with the training necessary to make them more effective at detecting, putting a cost to, and developing a plan to eliminate waste. Once workers realize that management sincerely wants and values their **input**, dramatic changes can occur. Work becomes more meaningful, and workers take pride in their work and in their organization.

Today, Dr. Deming is heralded as a genius, in part because he saw that Western organizations were managing the wrong way. They were working on controlling effects and not causes of the effects. They were trying to inspect quality of a product or service instead of ensuring that things were done correctly each step of the way. They were inspecting for defects instead of preventing defects. The defects are caused by abnormal variation in the system. Management was asking workers to do a better job in a system that had too much variation, instead of asking them to use their talent to identify the sources of variation in the system and develop a plan to minimize the variation.

Types of Variation in Processes

Variation is the result of many factors that constantly affect any process in manufacturing systems. Variation is inevitable because of random fluctuations and inconsistencies in the four Ms:

- Machines
- Materials
- Manpower
- Methods (procedures)

No two products, batches, or lots produced are exactly alike. The goal of statistical quality control (SQC) is to find out which variations are due to normal random fluctuations and which variations have an abnormal cause that can be detected and eliminated. In 1924, Dr. Walter Shewhart of Bell Telephone Laboratories developed the new paradigm for managing variation. As part of this paradigm, he identified two causes of variation:

- *Common cause variation* is inherent in a process over time. It affects every outcome of the process and everyone working in the process. Managing common cause variation thus requires redesigning the process.

● *Special cause variation* arises because of unusual circumstances and is not an inherent part of a process. Managing this kind of variation involves locating and removing the special cause from the process.

Chance or Common Cause Variation

When variation is random and stable, the process is in a state of statistical control. The process is operating according to the way it has been designed, and its performance is predictable. The stable pattern of variation is caused by conditions that are inherent to the designed manufacturing system. This is sometimes called **chance**, **normal**, or **common cause variation**. This type of variation is not a problem because it is normal to the system, sort of like background noise. It is constant and always there. An example is shown in **Figure 16.1.**

Special or Assignable Cause Variation

When variation is sporadic and unstable, the process is out of statistical control and no longer stable and predictable. More than likely the process is either making off-specification product or about to, and utilities and manpower are being wasted making slop instead of product. The unstable pattern is caused by such events as failing instruments, change in amount of catalyst, bad process feed, or human error. This is called **special**, **abnormal**, or **assignable cause variation** and is shown in **Figure 16.2.** Notice the large spikes above and below the normal variation baseline. Something special and not normal to the process caused these spikes.

No company wants to make a habit of making off-specification product because of the severe effect on profits. An ideal way to operate is to make product the way the customer

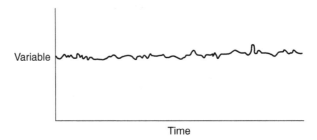

Figure 16.1   Normal Variation

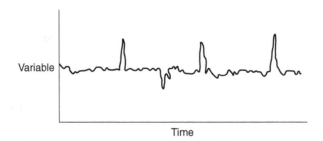

Figure 16.2   Special Cause Variation

wants it, and to continue to make the product that way every time. Though this sounds simple, it is not that simple. This consistency can be achieved but requires total quality management and lots of statistical quality control (SQC).

To practice SQC, a company should start with the same raw materials and process them the same way time after time. In other words, to ensure that their products will be in statistical control, they must have their processes in statistical control. The process variables (temperature, pressure, flow, level, composition) that impact the product properties must be monitored, and the variation must be constantly minimized. Consistency 24/7/365 is the operating mantra. Monitoring and minimizing the important variables in a process is the goal of SPC.

Now, think about a few very important facts about the advantages, and also the consequences, of achieving statistical control. If a process is operated in a state of statistical control, then:

- The process is doing what it was designed to do. It is doing the best it can.
- Operators are not able to improve the process.

Think about that last item. Operators can save their process unit money by eliminating waste and working smarter, but they have not improved the production process. If the process unit was designed to produce 50,000 pounds of 98 percent pure product in a 24-hour cycle, then that is the best it can do unless management redesigns the process and installs new equipment.

If the system displays statistical control and is still too variable to meet the needs of new customers who want a product purity of 98.4 percent, management has the responsibility of providing the means to improve the process. The operator cannot do anything about the inherent variability in the unit. The process and unit are working to designed engineering specifications. If the process must be improved, a major redesign may be necessary. Statistics and SPC might be able to help redesign the unit.

PROCESS VARIATION AS A TOOL

Every process always has variation. Provided workers use measuring instruments that are accurate enough, they will always find differences. If differences are not seen, they are not looking closely enough, or the measuring instruments aren't sensitive enough. (Remember the engineers who thought the test instruments were defective in testing the Ford transmissions in Chapter 3?)

So, besides being the source of waste, variation is the technical tool used to find problems, errors, complexities, and waste. Think of variation as a collection of clues that helps solve a problem; in this case, the problem is waste. Variation leaves tracks that employees can follow to identify the cause of variation and then reduce or remove it. Workers can gather data on the variation and turn that data into useful information by using several technical tools (see Chapters 14 and 15). The whole system of continuous improvement is based on using variation as a quality improvement tool. The workers can strive to reduce the range of variation.

Variation should be thought of as a tool. Analyzing process variation is the first step in meeting customers' needs. If an organization understands customers' requirements and reduces them to operational definitions, it can use its analysis of its own work processes to determine whether meeting the customers' needs is possible. If the present process cannot turn out products and/or services that meet customers' requirements, a study of the process variation can help the organization change the system to improve the process.

Employees must analyze and record the variation in the finished product, the work process, and the waste. They cannot see variation in a process unless they get close to it. The more the variation is studied, the more will be learned about how to improve the process and how to eliminate waste. When employees record the data and use statistical methods to convert the data to information, they learn what to eliminate and what to improve. Statistics requires measurement. Something that can't be measured is difficult to improve. Employees need to understand the work and work processes thoroughly to determine what to measure and how to use the measurements. They need to become masters of variation instead of victims of variation.

Operating Consistency

Operating consistency means performing job activities according to standard operating procedures or, in effect, doing the same thing the same way every time. A lack of consistency can become very expensive very quickly as time and resources are wasted seeking to return to consistency. Operating consistency can control variation that either *man* or *method* may add into the process. Every time an error occurs, someone must ask the following questions:

- Does a standard exist?
- Did we follow the standard?
- Is the standard adequate?

Once a company has established standards (procedures) and has ensured the standards are adequate and that employees follow them, then a company can claim that its processes are standardized and stable. The standard must be documented (written down). Oral (nondocumented) standards are subject to misinterpretation and gradual evolution. Remember that one of the key concepts of ISO registration was that a company would have a *documented* (written) quality process. Documentation provides a means for reliably capturing the standards needed for operating consistency.

VARIATION AND ACHIEVING STATISTICAL CONTROL

Variation in repetitive operations is not haphazard because it follows discernible patterns and obeys laws of probability that are predictable. The discovery, identification, and use of these basic variation patterns are a main concern of SQC. The object of an SQC effort is to reduce nonrandom variation to optimum levels, and to assess and control acceptable random variation. As previously mentioned, variation is present everywhere. We have all heard that no two snowflakes are alike, and that even identical twins have different fingerprints.

SQC helps an organization and its employees to understand variation and to minimize it in manufacturing systems. However, variability cannot be inspected, analyzed, or charted out of processes and products. Some process variation, called normal variation, will always exist. This is variation that is expected to be in the process; it is normal for it to be there, hence the name normal variation. Normal variation is due to the tolerance of the equipment. The following are two examples of normal variation:

● A centrifugal pump is rated for 600 ± 4 gallons per minute. In other words, the tolerance for the flow rate for this pump has a variation from 596 gallons to 604 gallons per minute.

● A temperature controller is rated to control temperature within 0.5°F of the displayed value. That is the normal variation of this controller.

These are examples of the normal variation of two pieces of process equipment. As long as the pump has an output of 600 ± 4 GPM, the control chart will reveal only normal variation in the flow. If the pump develops bearing problems or is damaged by cavitation, then its flow may drop to 575 GPM. This will show up on the control chart as abnormal variation. When a control chart reveals abnormal variation in a process, workers must take action to bring the variation back into the normal range.

Quality is achieved by reducing variation. Inspection or quality control cannot ensure quality because both are ineffective and expensive. These are postmortem evaluations, after the work is done. These methods occur too late to change the quality or lack of quality created.

SHEWHART AND DEMING ON VARIATION
Both Shewhart and Deming are recognized as extremely important figures in the quality revolution. Both had their own views on variation in a process and how to remove it.

Shewhart's Contribution to Controlling Variation
Walter Shewhart looked at process variability as being either within the limits set by chance or outside those limits. If variability was outside, he believed that the source of variability could be identified. He created the following distinction in variability (discussed earlier in this chapter):

● *Controlled variation* is attributed to *chance* and is characterized by a stable and consistent pattern of normal variation over time. It is due to the tolerances in equipment and materials.

● *Uncontrolled variation* is characterized by a pattern of variation that changes over time and can be attributed to an *assignable cause*. This variation is caused by a failure of something in the system or off-specification materials entering the system.

As an example, a worker in a manufacturing process making a series of discrete parts with a measurable dimension periodically selects and measures some of these parts. The measurements vary because the materials, machines, operators, and methods all inherently produce some variation. Such chance variation is consistent over time because it is the result of many contributing factors. Shewhart called these factors chance causes and thought of the resulting variation as **controlled variation**.

In addition to chance causes, occasionally special factors will have a large impact on the finished product. These factors might be machines out of adjustment, different operating crews using different procedures, or a change in raw materials. Shewhart said such factors could be identified and that the impact of such assignable causes would create a noticeable change

211

in the pattern of variation. He called this noticeable change in variation **uncontrolled variation**.

Shewhart's classification resulted in two different ways to improve any production process. If a process displays controlled variation, it is thought of as stable and consistent, and the variation present in the process is considered normal to the process. Therefore, to reduce the variation, the process itself must be changed (equipment improved or upgraded). However, if a process displays uncontrolled variation and keeps changing over time, it is both inconsistent and unstable. The way to improve this process is by identifying and removing the assignable causes responsible for the excessive variation. These two approaches to process improvement are fundamentally different:

- One looks for modifications to a consistent process (reduces normal variation).
- The other seeks to create a consistent process (removes assignable cause).

The approach to use depends upon the type of variation that the process displays. This means that the first step in attempting to improve a process is to determine if the process displays uncontrolled variation. The tool for detecting uncontrolled variation is Shewhart's control chart. His control charts, based on the laws of probability and statistics, are effective at detecting the presence of uncontrolled variation in any process. Shewhart published his first control chart in 1924, and then devoted the rest of his career to applying the principles of probability and statistics for achieving improved production processes.

Deming's Contribution to Controlling Variation

W. Edwards Deming had worked with Walter Shewhart at Western Electric and quickly realized the power of Shewhart's techniques. Deming wanted American industries to make greater use of control charts in their manufacturing processes. When the United States entered into World War II, Deming recognized an opportunity to push for the use of statistical methods in the manufacture of war materiel. He helped organize and teach short courses for defense contractors at Stanford University. By 1945, several thousand engineers and technicians had been given elementary training in statistical manufacturing methods.

When Deming went to Japan in 1947, he reformulated Shewhart's classifications of variation (assignable causes and chance causes). Rather than use these terms, which emphasized the source of the variation, Deming used terms that focused attention on who was responsible for doing something about the variation. He used the terms *special causes* and *common causes*.

Common causes of variation in a production process are variations that exist because of the design of the process (equipment tolerances) or the way that the system is managed. Because common causes are part of the system, they are the responsibility of those who control the system, the managers. Common causes of variation can only be reduced through action by management. Special causes of variation are localized and often specific to a certain worker or machine and can be identified and removed by action at the worker level.

Deming believed in the power of control charts. Management has to study the process and constantly improve products by finding and eliminating the sources of variation. Control charts are a way to pinpoint the sources of variation. As the variation in the process is reduced, the parts will be more nearly alike and the products will work better.

The Engineering Concept of Variation

A third concept of variation that has not yet been discussed is the **engineering concept of variation**. The engineering concept of variation and Shewhart's concept of variation have nothing in common because they have different objectives and yield different results. The engineering concept of variation has the object of just meeting specifications, resulting in *products that vary as much as possible,* because anything within specification is good enough. This concept does not seek to reduce or remove variation in a process. A few industries in America still operate this way.

Shewhart's concept is to remove variability, and the results are products that are as consistent as possible. The constant removal of variability results in continuous improvement of the process. Management has been trying conformance to specifications since the beginning of the Industrial Revolution, and has failed to achieve that goal for two hundred years. However, by applying Shewhart's concepts, the Japanese demonstrated the effectiveness of continuous process improvement by incrementally improving quality, increasing productivity, and gaining market share.

Again, remember the Ford transmissions in the Taguchi section of Chapter 3? That was an example of manufacturing to engineering specifications versus manufacturing to continuously remove variation and approach perfection.

THE NECESSITY OF CONTROL CHARTS

An ideal process can produce 100 percent conforming product and be in statistical control, but it will not always remain in the ideal state. A universal force called **entropy** acts on every process. Entropy is the gradual decay and decline of things and is similar to the aging process on the human body. Entropy causes an inevitable migration toward failure of all processes, including equipment and piping deterioration, decay, breakdowns, and failures. **Figure 16.3** is a control chart of a centrifugal wastewater pump. Note the gradual trend

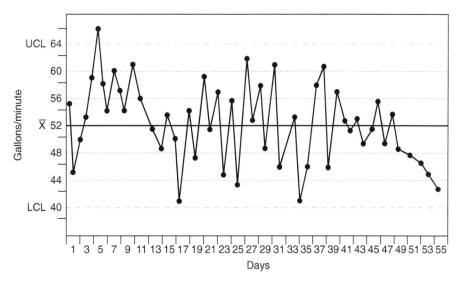

Figure 16.3   Entropy Revealed by a Control Chart

downward of values on the right-hand side of the chart. Prior to that, values had crossed back and forth between the centerline. This downward trend may indicate that the pump is damaged or failing. This is one reason the process industry has planned turnarounds (repair cycles). The only way this gradual failure of equipment and piping can be overcome is by continually detecting and repairing the effects of entropy. This means that the effects of a gradually failing process must be noticed before it fails. With such knowledge, the repairs can be made before failure occurs.

Every manufacturer must be able to identify two factors that adversely affect their process:

- The effects of entropy—the gradual failing of process equipment and instruments
- The presence of assignable causes—causes not normal to the process

Manufacturers can only meet the dual objectives of overcoming assignable cause and counteracting the effects of entropy by using process control charts, which track and measure variation on a continuous basis. No other tool provides such consistent and reliable information in a clear and understandable form. Because of this, any process operated without process control charts will eventually produce nonconforming product.

SUMMARY

Variation in repetitive operations follows discernible patterns and laws of probability that are predictable. The discovery, identification, and use of these basic variation patterns are a main concern of statistical quality control (SQC). The object of an SQC effort is to reduce nonrandom variation to optimum levels, and to assess and control acceptable random variation.

Variation can be tracked to reduce waste and improve a process. Variation is inevitable because of random fluctuations and inconsistencies in (1) machines, (2) materials, (3) manpower, (4) and methods. No two products, batches, or lots produced are exactly alike. The goal of SQC is to find out which variations are due to *natural* random fluctuations and which variations have an *abnormal cause* that can be eliminated. When variation is random and stable, the process is in a state of statistical control. A stable process can be kept consistently stable with control charts.

REVIEW QUESTIONS

1. Define *variation*.

2. Explain how statistical quality control (SQC) can assist in manufacturing systems.

3. List the four input factors in processes that contribute to random fluctuations and inconsistencies.

4. Explain the difference between *common cause* and *special cause* variation.

5. Monitoring and minimizing the important variables in a process is the goal of
   _____.

6. List two known facts about a process that is in statistical control.

7. Explain the two things a process technician should understand about variation.

8. Define *statistics,* and list several ways people use statistics on a daily basis.

9. Explain how *data* relates to *statistics.*

10. Explain the *engineering concept* of variation.

11. (True or False) An operator is working on a unit that stays in statistical control; however, some off-specification product is occasionally produced. The operator can help improve the process.

12. Describe Shewhart's two approaches to process improvement.

13. Describe the basic difference between Shewhart and Deming regarding variation.

14. Explain the necessity of control charts for process improvement.

GROUP ACTIVITIES

Divide into small groups and choose one of the activities below. Be prepared to report your group's conclusion to the class.

1. Each member of the group is involved in the process of driving to work (or college) everyday. Whether you get to work or college early or late is determined by the common and special causes of variance in this process. Make a list of the common causes (things that happen every day) and special causes (things that may happen only once or twice a month) that affect your ability to arrive on time to work or college. Explain why each is a common or special cause.

2. Pick two or three examples of real-life work processes you are or have been involved in, and list the common cause and special cause variation involved in each. Ask yourself the following questions, and explain your process and variations to the class:

   - Does a standard exist?
   - Did we follow the standard?
   - Is the standard adequate?

3. Coin toss process experiment: Place an open plastic bowl on the floor against a wall and give a student ten pennies. Ask the student to toss each penny one at a time into the bowl from a certain distance. Assign the student a certain place to stand before she begins the penny toss. Give her a set of written instructions that tell her how to toss the penny. Have several students do this. Ask the students as they complete the "process" why they missed so many or threw a coin so far off the mark, especially because it was simple process and they had clear instructions.

CHAPTER 17

Statistical Process Control

Learning Objectives

After completing this chapter, you should be able to:

- *Construct a histogram.*

- *Explain why operators would use a histogram.*

- *State the percentage of area under a normal distribution curve for $\pm 1\sigma$, $\pm 2\sigma$, and $\pm 3\sigma$.*

- *State the percentage of chance that a data point will fall within $\pm 3\sigma$.*

- *State the percentage of chance that a data point will fall outside of $\pm 3\sigma$.*

- *Explain the importance of the bell-shaped curve.*

- *Explain what a non-normally distributed population reveals.*

INTRODUCTION

This chapter and the next will review statistical methods that quantify variability in a data set. The idea is to simplify and focus on the characteristics that can answer questions like:

1. Where are the values centered (what is the average)?
2. How spread out are the values (what is the range of values)?
3. Are there any changes in the values over time?

Tabular displays of data (a page full of numbers) often do not provide a ready appreciation of many of the important characteristics of a collection of data. Remember, Chapter 15 pointed out that histograms or some type of chart best display tabular data. Large numbers of observations are best displayed in a frequency table or histogram that present a distribution in pictorial form. This distribution will yield information about the manufacturing process or will be used with another quality tool that will yield more information about the manufacturing process.

THE UNITED STATES AND THE BEGINNING OF STATISTICAL PROCESS CONTROL

"You don't have to be a genius to see a failure coming," meaning statistics can reveal a failing system. Neither a genius nor a crystal ball are necessary for predicting failure. Before launching into a discussion of **statistical process control (SPC)**, one of the cornerstones of a successful continuous improvement concept, briefly consider earlier discussions of the prevention versus detection quality management philosophies. Although Chapter 2 exposed some of the history of this information, the information in this chapter reveals the history of statistical process control in the United States.

Driving Blindfolded

Recall that one approach to quality management involves building products and then inspecting them to sort the good from the bad. Relying on inspection is the classical detection quality management approach. The prevention approach seeks to prevent defects from occurring. Though many tools are available for preventing defects, one of the most powerful is SPC. SPC places the responsibility for quality squarely in the hands of operators and not those of a downstream inspector who looks at the product after it is built.

Consider an analogy about driving an automobile on the freeway. You know how fast to drive to keep up with the traffic flow, and how to steer the automobile to keep it in your lane. The process of driving requires that we know the boundaries in front of, behind, and to the left and right of the car. We don't need anyone else in the car to tell us when to slow down or speed up, or to stay closer to the left or right side of the lane. We know how to keep the car in control. These skills occur almost automatically for those of us who have been driving awhile.

Now, imagine driving the same automobile on the same freeway and at the same speeds. However, this time you are blindfolded, and a fellow passenger in the car tells you when to speed up, slow down, veer left, or veer right. Could you drive a car this way? Probably, but it would be a terrifying experience. Taking it a step further, imagine forbidding the passenger (the one telling you to steer left or right and how fast to drive) from telling you anything. Driving in this manner guarantees an accident.

Now, think about how companies that rely on inspection as a means of assuring quality operate. The people building a product don't know if the widget they are building is good or bad (they are blindfolded as far as this information is concerned). After the blindfolded operators build a product, the product goes to an inspector who determines if it is good or bad. This is similar to you relying on a passenger for driving feedback, but with the added restriction that you would only receive feedback after leaving your lane or crashing into another car. In this example, relying on inspection is clearly an inefficient means of creating a quality product. Wouldn't it be better to take the blindfolds off of the operators and allow

Table 17.1 Statistical Quality Control

Statistical	With the help of numerical data,
Quality	we study the characteristics of the process,
Control	in order to make it behave the way we want it to behave.

them to evaluate the products as they build them? The answer is *yes*. SPC is the tool that removes the operators' blindfolds and moves organizations from a detection-oriented quality management philosophy to a prevention-oriented philosophy.

Understanding the variability in the data set from the process will help employees better understand the process and reduce variability. This involves using a statistic to control the process (see **Table 17.1**).

From the Industrial Revolution to the Information Age

Many of us, when faced with examples of poor quality, often hear comments such as "There's just no pride of workmanship anymore." The implication of this comment is that quality deficiencies are often due to workmanship or operator errors. However, modern quality management maintains that very few nonconformances are due to worker error; instead, most are due to defective manufacturing processes, poor management systems, and poor product design.

Early in our nation's history, craftsmen individually made most items. A single individual produced a chair, for example, turning raw lumber into a finished product. Clothes, guns, wagons, and virtually every manufactured product were produced the same way. Techniques of mass production had not yet arrived, so a single individual did in fact have responsibility for building an item from start to finish. Any mistakes in its fabrication were exclusively that individual's. Craftspeople recognized this, and frequently took great care in creating an item, and did indeed have "pride of workmanship."

The beginnings of change emerged in the 1920s. Walter Shewhart had done some work in designing a headset for U.S. Army radio operators. After having the heads of 10,000 soldiers measured, he noticed some interesting statistics. Shewhart continued his work with statistics at Bell Laboratories and recognized that, just as people's head sizes varied, so too did the dimensions of items manufactured using mass production techniques. Shewhart recognized that the variability associated with manufactured component dimensions—length, width, weight, height—followed a normal distribution, and he reasoned that one could track component dimensions to determine if they were starting to drift out of the normal range. Shewhart recognized that, as long as the dimensions remained in the normal range, their variability was under control. If they started to drift out of the normal range, something special was occurring to take the process out of control. If one could see this departure beginning to emerge by tracking dimensions, one could then introduce corrections early and prevent making any parts that were outside allowed tolerance limits. Shewhart refined his approach and published the book *Economic Control of Quality of Manufactured Product* in 1931, but manufacturing management wasn't ready for it.

Statistical Process Control Helps the War Effort

The United States entered World War II in 1941. All kinds of complex items were suddenly being made in enormous quantities to support the war effort. This was particularly true in the ordnance industry, where bombs and bullets were being made by the millions. A few insightful people recognized that simply sampling and performing inspections on munition lot samples would not work under these circumstances. The cost of rejected lots of ammunition would be enormous in financial terms, but trivial compared to the cost of not providing ammunition—or even worse, providing munitions that did not work—to military forces that desperately needed it. Clearly, the United States needed a management approach that would reduce scrap levels and ensure that only good products reached the troops. A preventive approach was needed. Walter Shewhart's work came to light again, and the United States adopted statistical process controls on a large scale for the first time. This first major application involved the wartime munitions manufacturing facilities in the United States. Sadly, when World War II ended, so did Americans' interest in statistical process control, and the concept died in the United States to a great extent.

Japan Accepts What the United States Rejects

As stated in Chapter 2, Japan was a devastated nation at the close of World War II. The Japanese industrial base was effectively destroyed, the government no longer functioned, and basic human needs could not be met. The United States installed General Douglas MacArthur as military governor of Japan immediately after the war. MacArthur recognized that one of his immediate priorities was to help the Japanese begin rebuilding their industrial capabilities. He enlisted the aid of an American management consultant named W. Edwards Deming. When Deming came to help the Japanese, he found a remarkable situation.

Japan is a small island nation with no natural resources except the Japanese people's intelligence and industriousness. The Japanese then, as now, were a frugal people. A management philosophy that required minimal waste was particularly attractive to a country that had to import virtually all its raw materials. Deming's philosophy, SPC, offered just such an approach. The Japanese eagerly embraced Deming's teachings. The results were not immediate, but Deming instilled in the Japanese an underlying statistical approach that would offer strong worldwide quality and marketing advantages in coming decades.

Oblivious to the emerging quality of Japanese products, American industry after World War II remained firmly entrenched in an inspect-and-detect quality management philosophy. Besides, anything made could be sold. Manufacturing management saw no need to minimize scrap or undertake seemingly complex SPC approaches. Why change when life was so good? Americans' first inklings of Japan's looming threat occurred when Japan began to make significant inroads into the American consumer electronics market in the mid- to late 1960s. Another early indication occurred when Honda began to dominate the international motorcycle market with its inexpensive, high-reliability, and high-quality motorcycles.

In 1973, something else happened that changed Americans. Oil embargoes hit the United States, and suddenly the gas-guzzling monsters (getting twelve miles to the gallon) that American automobile manufacturers had offered for decades began to lose market shares to smaller and more economical foreign automobiles. Cars like Hondas, Toyotas, and Datsuns impressed American consumers with their quality, reliability, and low cost. The oil embargoes ended but Americans had experienced high quality and weren't about to give it up.

What did the Japanese have in automobiles, consumer electronics, cameras, and motorcycles that American industry did not? In one word: *quality*. American consumers recognized that Japanese products more nearly met their needs and expectations. The fit, finish, and inherent quality in Japanese products were irresistible. SPC was (and still is) one of the key Japanese management tools for attaining these attributes.

The United States Awakens

American industry began to recognize the extent of its problem at the start of the 1980s. Automobile manufacturers had lost major portions of their market share to the Japanese. Harley-Davidson, the only remaining American motorcycle manufacturer, had become a minor player in an industry it had previously dominated. Most of the American camera market belonged to the Japanese. Most American consumer electronics companies that once had dominated their markets either lost significant market shares or ceased to exist.

An ironic phenomenon occurred when American industry began the struggle to turn the quality situation around. The Japanese had previously turned to the United States for industrial management guidance, but Americans now sent people to Japan to study why Japanese manufacturing methods worked better. Americans found that one of the dominant reasons for Japan's quality superiority was their use of statistical methods for controlling manufacturing processes.

HISTOGRAMS (REVISITED)

As introduced in Chapter 15, a histogram is a statistical method that makes bar charts of the values from a data set grouped into classes. Histograms are important diagnostic tools because they give a clear picture of data distributions that might otherwise be difficult to visualize. When the intervals for the histogram are chosen and the histogram bar graph plotted, the characteristic bell-shaped distribution should occur if a process exhibits a normal distribution (bell-shaped distributions will be discussed in the next few pages). A histogram should always be constructed before calculating other statistics because they help detect non-normal distributions. Non-normal distributions reveal special cause in a population.

Table 17.2 presents condenser temperature data from which to construct a histogram. No patterns stand out in the raw data shown in Table 17.2. This information—patterns—will be revealed by using the histogram quality tool. Use the following steps to construct a histogram:

1. Determine the appropriate number of data classes (intervals) for the 60 measurements. You can do this in several ways, but we will use the method that takes the square root of the sum of the observations. The square root of 60 is about 7.8. This means we can use either seven or eight data classes (intervals).

2. Group the data into equal size classes. For convenience, because the condenser has eight different temperature values, we'll divide the data into eight classes with a range of 1°F. If the condenser showed sixteen different temperature values, we would use an interval of every two degrees.

Table 17.2 Condenser Temperature Data (°F)

Day	Temp 1	Temp 2	Temp 3	Day	Temp 1	Temp 2	Temp 3
1	49	50	50	11	48	50	52
2	52	48	53	12	47	48	50
3	50	51	51	13	49	54	53
4	50	50	50	14	52	49	54
5	49	51	51	15	54	51	51
6	52	50	47	16	50	48	50
7	51	52	53	17	50	51	52
8	48	52	51	18	49	52	48
9	50	51	52	19	50	49	49
10	53	51	49	20	51	50	52

Note: Readings were taken at 8:00 A.M., 4:00 P.M., and 12:00 A.M.

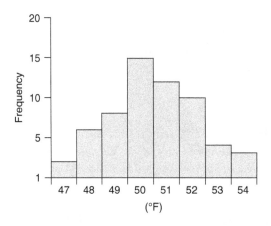

Figure 17.1 Histogram of Condenser Temperature Ranges

3. Begin with the smallest temperature values on the left and proceed with increasing temperatures to the right.

4. Construct the histogram with the y-axis containing the number of measurements in each class and the x-axis containing the midpoint of each class interval. The histogram will look like that in **Figure 17.1.**

Histograms allow a serious "peek" at a population. They give good estimates of what populations look like (bell-shaped, bimodal, skewed, and so on), how much spread there is, and where the population is centered. That's why histograms are such powerful tools. Notice that the histogram yielded the most common temperatures to be between 50°F and 51°F, and a spread between 47°F and 54°F. Also, the histogram revealed a critical piece of information: a continuous line drawn across the tops of each bar reveals, roughly, a bell-shaped curve (see **Figure 17.2**), also called the normal distribution. Normal distributions are necessary for the creation of statistical process control charts.

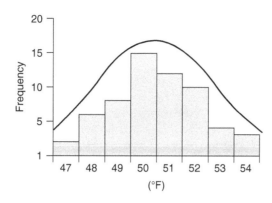

Figure 17.2   Bell-Shaped Curve of Condenser Histogram

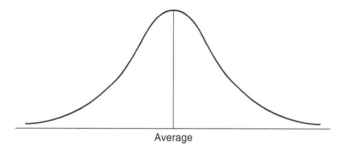

Figure 17.3   Normal Distribution (Bell-Shaped) Curve

THE NORMAL DISTRIBUTION

Processes that are stable and predictable have a fixed pattern of variation that can be described by a mathematical curve. This predictable pattern is known as the ***normal distribution*** and is like a histogram with two special features:

- First, the normal distribution is a smooth curve.
- Second, the normal distribution has a particular shape that resembles a bell and is often called a bell curve (see **Figure 17.3**).

The average value of a normal distribution is always at the center, and there is a quantity called **standard deviation**, designated by the Greek letter *sigma* (σ) associated with it. An equation exists to calculate σ for a group of data points, but it will not have to be used in this text. The important thing is to understand what standard deviation means. The standard deviation is the measure of the dispersion of observed values from the mean (how far away values are from the centerline). The standard deviation is important to determine because the larger the standard deviation value, *the larger the variation in the process*. The larger the variation in the process, the more waste attributed to the process.

To better understand standard deviation, divide the normal distribution curve into equal parts in the horizontal direction (see **Figure 17.4**). The horizontal distance from the average to the first line is called one standard deviation ($+1\sigma$). The horizontal distance from the average to the second line is called two standard deviations ($+2\sigma$), and the distance from the average to the third line is three standard deviations ($+3\sigma$). If the standard deviations are to the right of the centerline, they are positive. If they are to the left, they are negative.

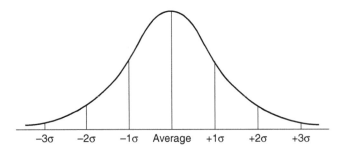

Figure 17.4 Standard Deviations under the Bell-Shaped Curve

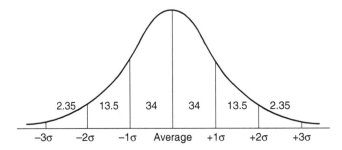

Figure 17.5 Area Percents under the Bell-Shaped Curve

Notice that one, two, and three standard deviations are indicated on both sides of the average (centerline). The equation for standard deviation is set up so that the area under the curve that covers the distance from the average to $+1\sigma$ is about 34 percent of the total area under the curve (see **Figure 17.5**). Thus, the area from one side of the average line and the other side of the average line ($\pm 1\sigma$) would be 68 percent of the total area under the curve ($2 \times 34 = 68$). In practical terms, this means that a normal distribution of 100 different measurements would have 68 percent (or 68) of the measurements falling within ± 1 standard deviation of the average.

To state that in a little easier way to understand, if we had 100 temperature readings for the condenser, 68 percent of them would fall within the range of $\pm 1\sigma$, or 68 readings would be within that distance from the average reading of 50.5°F. Thus, a large number of the readings will be around the centerline (average).

Now, look at the area under the curve from $+1\sigma$ to $+2\sigma$ (as shown in Figure 17.5). It has about 13.5 percent of the whole area under the curve, and the area from $+2\sigma$ to $+3\sigma$ is 2.35 percent of the whole area under the curve. As noted earlier, 68 percent of the measurements will fall between $\pm 1\sigma$. Now, adding the areas $13.5 + 34 + 34 + 13.5$ yields 95 percent. This means that 95 percent of all the measurements will fall between $\pm 2\sigma$. Lastly, add the area under the curve that represents $\pm 3\sigma$ to the other total. Adding $2.35 + 13.5 + 34 + 34 + 13.5 + 2.35$ yields 99.7 percent, meaning 99.7 percent of the 100 condenser measurements (99.7) will fall between $\pm 3\sigma$ if the distribution is normal.

Thus, the normal distribution statistic reveals a 99.7 percent certainty that a measurement will fall between $\pm 3\sigma$ when a normal distribution is present. Remember, a normal distribution means a stable process is running according to design. Thus, 99.7 percent is a pretty

224

good certainty and the kind of reliability manufacturers look for in their processes. There is only a 0.30 percent chance that a measurement will be outside of $\pm 3\sigma$. To put it in terms of good or bad product, it means 3 out of 1,000 widgets or 3,000 out of 1,000,000 widgets will be bad.

The point of all the averaging and range finding is to determine if the raw data are normally distributed. If the data aren't, it makes no sense to refer to a standard deviation of the raw data. Is the standard deviation that important? Yes. When making control charts, formulae use factors for calculating the upper control limits (UCLs) and lower control limits (LCLs). The factors D_3, D_4, and A_2 are factors that fix the UCL and LCL approximately three standard deviations above and below the centerline. This will be discussed in more detail in Chapter 18.

The shape of a histogram often provides clues to the causes of a problem if it does not form a normal curve. **Figure 17.6** shows curves of populations that are not normally distributed. Histogram 1 is of golfing scores and is skewed to the right. This histogram reveals professional golfers' scores (the higher scores on the right) were mixed with those of duffers; in effect, it displays two populations. Histogram 2 is bimodal, which also reveals that two processes (populations) are involved. This could be two different machines calibrated differently or two different shifts operating differently. Whenever two different populations appear in the same data group, we will usually not have a perfectly normal curve, and hence, the population cannot be used for SPC charts.

It is important to have a normal distribution in SPC work. However, if the data are not normally distributed, a statistical tool called the **central limit theorem** allows the statistician to obtain a normal distribution. In preparing control charts, data are arranged

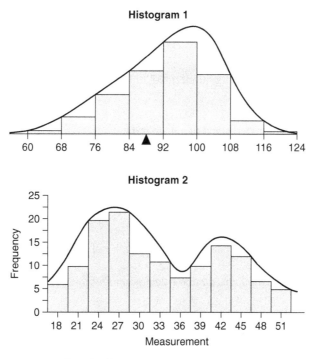

Figure 17.6   Non-Normally Distributed Populations

in groups called subgroups, and averages are determined. This can be done because the central limit theorem states that we will tend to have normal or nearly normal distributions when we work with averages. Therefore, it is possible to get normal distributions even if the data are not normally distributed.

SUMMARY

This chapter presents a brief history of the steps that led Americans to accept statistical process control. It also is an introduction to SPC and focuses on statistics that answer the following questions:

- Where are the values centered (what is the average)?
- How spread out are the values (what is the range of values)?
- Are there any changes in the values over time?

Tabular displays of data often do not provide a ready appreciation of many of the important characteristics of a collection of data. Large numbers of observations are best displayed in a frequency table or histogram that present a distribution in pictorial form. Histograms are bar charts of the values from a data set grouped into classes. Histograms give good estimates of what populations look like (shape), how much spread there is, and where they are centered.

Processes that are stable and predictable have a fixed pattern of variation that can be described by a mathematical curve. This predictable pattern is known as the normal distribution; it is a smooth curve that is bell-shaped and is thus often called a bell curve. The normal distribution statistics reveal a 99.7 percent certainty that a measurement will fall between $\pm 3\sigma$ when a normal distribution is present.

REVIEW QUESTIONS

1. How is an operator "driving blindfolded" if his work process relies on inspection to determine quality?

2. List the information contained in a histogram.

3. Define *standard deviation*.

4. State the percentage of area under a normal distribution curve for $\pm 1\sigma$, $\pm 2\sigma$, and $\pm 3\sigma$.

5. State the percentage of chance that a data point will fall within $\pm 3\sigma$.

6. State the percentage of chance that a data point will fall outside of $\pm 3\sigma$.

7. Explain why a normal distribution is important for a process that will use statistical process control.

8. Explain what a non-normally distributed population reveals.

GROUP ACTIVITIES

Divide into small groups and do both of the activities below.

1. A customer complained to a supplier that some of the barges delivering feedstock to one of its plants were arriving later than normal. The customer was concerned that this tardiness might worsen and eventually interfere with his company's production schedule. In the past, barged feedstock transit time rarely exceeded eight days. Use the following data to (1) make a histogram, (2) determine if the customer has a valid complaint, and (3) determine what percentage of barges is taking longer than eight days to arrive.

   A total of thirty-five barges delivered feedstock to the concerned customer. The following are the transit times (in days) for each barge:

   5, 10, 6, 6, 7, 8, 7, 8, 9, 5, 9, 6, 6, 5, 8, 8, 9, 7, 9, 6, 10, 5, 6, 7, 8, 6, 8, 7, 9, 7, 10, 8, 7, 7, 8

2. A company has a stable process in statistical control. It produces 2,500 items a day. How many of those items will fall within $\pm 1\sigma$? How many will fall within $\pm 3\sigma$? How many will be defective?

CHAPTER 18

SPC and Control Charts

Learning Objectives

After completing this chapter, you should be able to:

- *Explain the usefulness of statistical process control (SPC) charts.*

- *List five advantages of SPC charts.*

- *Construct an $\bar{X}$ chart.*

- *Explain what information an $\bar{X}$ chart reveals that an R chart does not.*

- *Explain what information an R chart reveals that an $\bar{X}$ chart does not.*

- *Construct an R chart.*

- *List the rules for determining whether a process is out of control even though all data points are within the $\pm 3\sigma$ limits.*

INTRODUCTION

Previous chapters have revealed how the introduction of mass manufacturing required the creation of some type of statistical information that monitors the manufacturing process. That information alerts operators when the process is trending toward production of off-specification product, or when special cause has entered the process. That particular statistic is Shewhart's control chart.

Many quality-conscious purchasers require potential suppliers to have fully operational SPC programs in place before they can become approved vendors. In refineries and many other places, almost all continuous processes use SPC routinely. Control charts work because variation in a stable process almost always follows a normal distribution that can be expressed mathematically.

The normal distribution has more measurements near the average and fewer measurements farther from the average. The control limits are set so that random variations will almost always fall within those limits (three standard deviations) in a normal distribution (99.7 percent of the time). This allows manufacturers to separate random (common) causes, which are part of the system, from special causes. The absence of special causes indicates the process is probably stable. A number of other tests can determine if the process is out of control, even if all points are within the control limits.

SPC AND CONTROL CHARTS

Combine the run chart and the histogram with a statistical formula, and you have the control chart. Control charts are probably the most powerful tools for controlling and improving repetitive processes. Walter Shewhart invented the control chart to improve the understanding of the variation in a process. Because the control chart is based on mathematical formulae that sort random from nonrandom events, it separates the causes of variation into two classes: those that are a part of the system (common or normal causes), and those that come from outside the system (abnormal or special causes).

Stabilizing a process is a significant achievement because people can determine the capabilities of a stable process and can count on the output of the process being within a certain range. Furthermore, stability provides a base from which the effects of changes in the system can be measured accurately. This stability is critical so that people can separate the random variation from the change caused by the trial improvement. Otherwise, they might be misled by random variation to conclude that the changes they made were either better or worse than they really were.

Advantages of SPC

Engineers using computers and statistical software create most control charts. Process technicians probably won't be asked to construct control charts, but they will be asked to monitor control charts and determine what they are indicating about the variation in the process. Some of the advantages of SPC are:

- Improved product quality
- Increased quality consciousness
- Cost reduction
- Data-based decisions
- Predictable processes

As stated earlier, companies in the process industry use statistical thinking and statistics to help them determine when to make changes to a process. Changes are not made unless proof exists to substantiate the change; otherwise, things are usually made worse. Some processes—such as a distillation tower—take many hours to return to normal when they become upset. This results in hours of off-specification material and wasted resources.

SPC is used to continuously monitor process performance with charts and the mathematical probabilities of the normal distribution. If a company accepts Crosby's definition of quality as conformance to requirements, then it must have a process capable of consistently meeting those requirements. It must use SPC to define the process capability and nonconformances, and to assist in eliminating the nonconformances. For quality improvement to occur, these concepts should be clearly understood and be a part of each individual's job.

"Doing it right the first time" and defect prevention are sensible goals; however, these goals cannot be reached without quantitative measures (data) of quality to effectively identify variability in a process. SPC and its associated problem-solving techniques contain the quantitative tools that allow this objective approach to quality. The purpose of a control chart is to (1) show reliability as a supplier to external customers, and (2) improve the process for internal customers by removing fluctuations and disturbances.

What is SPC?

SPC is a technique of monitoring, measuring, and controlling the performance of a process using charts and graphs. The SPC technique recognizes that variations will always be in a process; however, as long as the variations are normal, there is no reason to worry. SPC divides variation into two major groups: normal variation and abnormal variation. Abnormal variation can be caused by a malfunctioning pump, a temperature controller out of calibration, bad feed quality, and so on. The cause doesn't have to be something major. As an example, during a reactor startup to make polyethylene, just a minor amount of impurity in the reactants can cause major variations in product quality.

SPC charts and associated problem-solving techniques provide a picture of the performance of a process. This picture can be analyzed to detect an incipient problem and make a correction before a process produces off-specification product. Further, the chart can be analyzed to help identify the root cause of a problem—the first step toward problem elimination and prevention. SPC will enable processes to be run more consistently and increase the percentage of product that meets customer specifications. It enables manufacturers to understand their processes and to answer the following questions: (1) Are we in control of the process? and (2) Is there room for improving the process (reducing variation)?

Training the Operators

SPC is part of total quality management and cannot be practiced in isolation. Those actually running the process must be involved, not just management. One of the most important elements of any SPC implementation program is training. The concepts involved in SPC are somewhat abstract and require an elementary understanding of the nature of statistical distributions. Training the operators who will be interpreting the control charts is critical because one of the underlying purposes of SPC is to place control and monitoring of the process in the operators' hands. If the operators do not understand the principles behind SPC, the process will be viewed as just another chore, and not a tool to help them produce quality product. With management's involvement and encouragement as well as operators trained in the basics of SPC, the manufacturing process is ready to be controlled by SPC, not inspection.

STATISTICS FOR SPC

Before looking at the various statistical methods used in SPC, we must understand some basic statistical principles. Earlier, when we looked at data collection, population was defined as all members or elements of a group (the entire 50,000 gallons in the storage tank, for example). A sample was a small portion of that population.

We can sample for **attribute data** (also called **discrete data**)—qualitative data indicating the number of items conforming or failing to conform to specifications. Examples of attribute data are good or bad product and the percentage of late shipments to customers. We can also sample for **variable data** (also called **continuous data**), which are data obtained by measuring variables such as flow rate, temperature, or pressure in a vessel.

Data will have a distribution pattern, that is, a pattern of variation. The following can describe distributions:

- **Central tendency**—depicts where the values are centered
- **Spread** (also called *range*)—reveals how spread out the values are

Most process industry distributions are, or approximate, a normal distribution. This is explained in more detail later in the chapter. In a normal distribution, the centerline is the average (or mean), and the spread is the **standard deviation**. The mean is also known as the arithmetical average and is represented by the symbol $(\overline{X})$.

A measure of central tendency (mean) reveals information about a data set but does not adequately characterize the data. It leaves something out. To learn more about the data, the variability (or spread) must also be measured. Spread can be characterized as range or as standard deviation. The range is the difference between the highest and lowest observed values and is represented by the symbol R. The standard deviation reveals how close the values in the sample are to the mean, and how far the data spread out from the mean. The standard deviation is represented by the symbol σ, which is a Greek symbol called **sigma**.

BASIC $\overline{X}$ AND R CONTROL CHARTS

One of the most useful kinds of control charts deals with data from variables that can be measured by numbers, such as flow, temperature, or pressure. Such control charts are called $\overline{X}$ and *R charts;* how they are constructed will be discussed in the following pages.

Control charts contain the following:

- The **centerline** reveals the process average and is represented by the symbol $\overline{\overline{X}}$.

- The **upper control limit (UCL)** is the farthest range above the centerline where a data point can be plotted and still be considered normal variation. The UCL is represented by $+3\sigma$. Data points falling above the UCL indicate the process is experiencing abnormal variation.

- The **lower control limit (LCL)** is the farthest range below the centerline where a data point can be plotted and still be considered normal variation. The LCL is represented by -3σ. Data points falling below the LCL indicate the process is experiencing abnormal variation.

When a control chart is constructed, the x-axis depicts time, and the y-axis depicts the variable (temperature, flow, and so on) or data plotted. Though there are several types of control charts, the one most used by the processing industry is the control chart of variables. The control chart of variables will be the only one discussed in this book; however, a good reference for the various types of control charts is McNeese and Klein's book, *Statistical Methods for the Process Industries*.

$\overline{X}$ and R charts are used for variable data. Though they are two separate charts, they are usually combined on one sheet of paper and paired up to reveal information. The $\overline{X}$ chart studies the variation in the subgroup averages, and the R chart studies the variation in the subgroup ranges. $\overline{X}$ and R charts are used when data are often available and can be subgrouped rationally, and when there is a need to detect differences in subgroups plotted over time.

How to Construct $\overline{X}$ and R Control Charts

Suppose, after a unit turnaround, a new, larger condenser has been installed on a distillation tower of the styrene unit. A crew of operators on the unit has been given the duty to track the temperature variation across each shift and to collect the data to create a control chart. They would take the following steps to construct a control chart.

Collect the Data As stated in an earlier chapter, collecting the right data and how the data are collected are important; otherwise, time and resources are being wasted. The unit engineer, management, and unit operators should get together and conduct the following steps:

1. Decide on the information needed. What changes and variations do they need to understand? Those that occur from day to day, hour to hour, or from shift to shift? In this case, management is interested in the condenser's temperature fluctuations over a twenty-four-hour period.

2. Select the **subgroup** size (n), which is typically between three and five. A subgroup is a collection of readings. A subgroup size of three was selected. In this case, the subgroup was the three daily temperature readings.

3. Select how frequently the data should be collected—every four hours, every twelve? The data should be collected and plotted *in order* of *production*. Data were collected every eight hours.

4. Wait until about twenty to thirty subgroups have been collected before calculating control limits. Twenty days of data were collected.

Data from **Table 18.1** are used to construct a control chart. Operators collected the data from the condenser temperature local instruments as they made their rounds.

Now the group creating the SPC chart will use the data sheet to conduct the following steps:

1. For each subgroup, record the individual sample result, and then calculate the subgroup average. The subgroup averages for days 1 through 20 are shown in **Table 18.2.** Values have been rounded off to the nearest tenth.

2. For each subgroup, calculate the range. The range is the difference between the maximum value and minimum value of the subgroup. This is also shown in Table 18.2.

Table 18.1 Condenser Temperature Data (°F)

Day	Temp 1	Temp 2	Temp 3	Day	Temp 1	Temp 2	Temp 3
1	49	50	50	11	48	50	52
2	52	48	53	12	47	48	50
3	50	51	51	13	49	54	53
4	50	50	50	14	52	49	54
5	49	51	51	15	54	51	51
6	52	50	47	16	50	48	50
7	51	52	53	17	50	51	52
8	48	52	51	18	49	52	48
9	50	51	52	19	50	49	49
10	53	51	49	20	51	50	52

Note: Readings were taken at 8:00 A.M., 4:00 P.M., and 12:00 A.M.

Table 18.2 Condenser Subgroup Averages

Day	Temperature Average	Range Average	Day	Temperature Average	Range Average
1	49.6	1	11	50	4
2	51	5	12	48.3	3
3	50.6	1	13	52	5
4	50	0	14	51.6	5
5	50.3	2	15	52	3
6	49.6	5	16	49.3	2
7	52	2	17	51	2
8	50.3	4	18	49.6	4
9	51	2	19	49.3	1
10	51	4	20	51	2

Constructing the Control Chart

1. Find the average range, R (called $\overline{R}$), by calculating the average of all the R values for all the subgroups. The value $\overline{R}$ will locate the centerline of the R chart and will also be used in additional calculations.

Average range = sum of subgroup average ranges divided by twenty.

$\overline{R} = 57/20 = 2.85$

Table 18.3   SPC Control Chart Constants

Number of Observations in Subgroup	Factors for Averages A_2	Factors for Ranges	
		D_3	D_4
2	1.880	0	3.267
3	1.023	0	2.575
4	0.729	0	2.282
5	0.577	0	2.114
6	0.483	0	2.004

2. Calculate the control limits for the R chart using the following formulae:

$UCL_R = D_4 R$ (D_4 is a control chart constant found in **Table 18.3** and is dependent on subgroup size.)

$LCL_R = D_3 R$ (D_3 is a control chart constant found in Table 18.3 and is dependent on subgroup size.) Sometimes, due to the small subgroup size (usually less than seven), the R chart has no lower control limit.

The calculations for the control limits for $\bar{R}$ are:

$UCL_R = D_4 R \rightarrow UCL_R = 2.575\ (2.85) \rightarrow 7.338$ rounded to 7.34

$LCL_R = D_3 R \rightarrow LCL_R = 0\ (2.85) \rightarrow 0.0$

We have completed the calculations for the R chart, and they reveal that:

- The average range value (centerline) for R is 2.85°F.
- The upper control limit (three standard deviations) for the range is 7.34°F.

As long as the temperature variability does not exceed 7.34°F, then 99.7 percent of the values will be within the R chart upper control limits, indicating that there is no excessive variability in the system and that it is operating as designed.

Constructing the X̄ Chart

1. Find the grand average, X (also called $\bar{\bar{X}}$). This is another very important calculated value because it locates the centerline of the X̄ chart. To calculate $\bar{\bar{X}}$, simply find the average of all the subgroup values on the condenser data sheet. The value $\bar{\bar{X}}$ is the average of all the averages. ($\bar{\bar{X}}$ should calculate out to be 50.475.)

2. Calculate the control limits for the X̄ chart using the following formulae. A_2 is a control chart constant found in Table 18.3.

$UCL_X = \bar{\bar{X}} + A_2 \bar{R} \rightarrow = 50.475 + 1.023\ (2.85) = 53.39 = 53.4°F$

$LCL_X = \bar{\bar{X}} - A_2 \bar{R} \rightarrow = 50.475 - 1.023\ (2.85) = 47.559 = 47.6°F$

3. Calculate the control limits for plus and minus one and two standard deviations by dividing the difference between the upper and lower control limits by six. This will give you the value of one standard deviation.

Constructing the Chart

1. Obtain a blank control chart. Choose an appropriate vertical scale for the x-axis and y-axis so that all X and R values can be included. Try to use as much of the graphing area as possible so that both charts are large and easy to read and record on.

2. Draw the upper control limit on the R chart as a heavy dotted line and label. Draw the centerline of the R chart as a solid line and label.

3. Draw the control limits of the $\overline{X}$ chart. Most charts show the limit of the plus or minus one and two standard deviations in dashed or colored lines, and the upper and lower control limits expressed as solid lines. Write the limit value beside each line. Draw the centerline as a thick solid line and label.

Plotting Data

1. Once you have created the control chart, you are ready to use the chart to monitor your condenser system. Plot the subgroup ranges of new data on the R chart and connect each plotted point with a straight line. See the example chart (**Figure 18.1**), which is partially filled out with data from Table 18.2.

2. You are also ready to plot the subgroup averages on the $\overline{X}$ chart and connect each plotted point with a straight line (see Figure 18.1).

3. When a chart becomes filled, you create a new control chart with the same control limits and continue plotting. (Later in this chapter, you will read about when it might be necessary to change the control limits for that process.)

INTERPRETING CONTROL CHARTS

Now, we have a control chart that reveals the data points of the stable process and its normal population of a variable. Plus, the chart reveals a lack of stability in the process (special cause) in two ways:

1. By points that fall above the UCLs and below the LCLs
2. By points that are not randomly distributed

Nonrandomly distributed data points will be discussed in detail near the end of this chapter. Points that fall outside of the upper or lower control limits and points not randomly distributed indicate that the process is unstable and that there is an identifiable cause for the variation.

Always study the $\overline{R}$ chart first. If the $\overline{R}$ chart shows excessive variability (too many points out of the control limits), the process is unstable and excessive variability must be removed. If the $\overline{R}$ chart shows excessive variability, then the control limits on the $\overline{X}$ chart have little meaning. It is important to realize that the $\overline{R}$ chart provides information about the variability of the system, and the $\overline{X}$ chart provides information about how average the system is. The system cannot be "average" if it has excessive variability. It is also important to

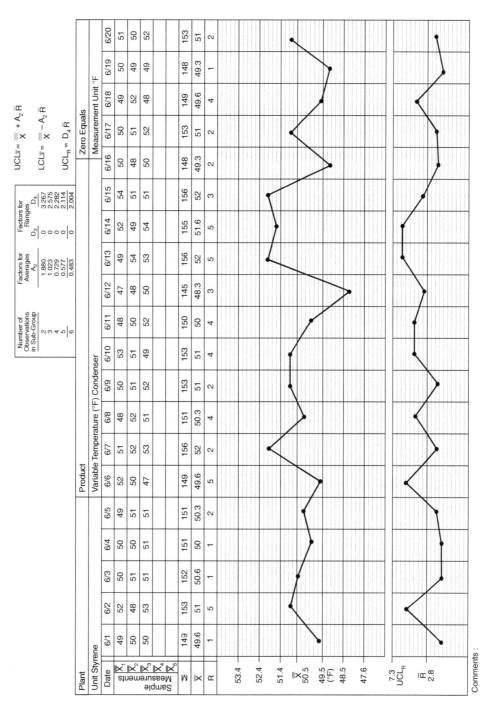

Figure 18.1 Control Chart for Tower Condenser

understand what "statistical control" really means. A state of **statistical control** is a state of randomness. The variations seen in a state of statistical control are due to chance alone, with nothing external disturbing the stability of the process.

Taking Action

Control charts reveal when a process is unstable and alerts manufacturers to take corrective action. They indicate when the process variation has become unacceptable. At that point, a diagnosis of the cause of the variation and corrective action is required. Control charts are worthless unless corrective action is taken when evidence of excessive variation occurs. The idea of statistical process control is based on the idea that someone will take action to improve the situation if process variation is unacceptable.

Recalculating Control Limits

As systems and processes improve through better equipment or controls, the control charts may reveal that the variability ($\overline{R}$ chart) or the averageness of the process ($\overline{X}$ chart) changes. This might suggest the need to recalculate new values for the centerlines and control limits. This should be done only when it needs to be done. If the process has definitely been changed such that the points on the control charts show a distinctly new pattern over an extended time, new centerlines and control limits should be calculated. Remember, control charts are to be constructed so that they represent the reality of what is happening as nearly as possible. The recalculation of new centerlines and control limits implies awareness that the process has changed (improved).

Figure 18.2 shows the reduction in variability as process equipment was upgraded and the process improved. Notice how the upper control limit has decreased from 1.037 inches to 1.025 inches, and the lower control limit has decreased from 0.937 inches to 0.962 inches. This is a reduction in variation of 63 percent, which resulted in a tremendous saving on scrap.

TESTS FOR LACK OF CONTROL

By now, you must realize that, when a data point falls outside of three sigma limits, the process has changed and is no longer in control. Special cause has changed the process population. However, the process can still be out of control even when all points are within the three sigma limits. For a normal distribution, *the distribution of points within three sigma limits is predictable and there are statistically based rules for evaluating and interpreting the distribution of sequential data points within the three sigma limits.* Sometimes, the data points fall in a

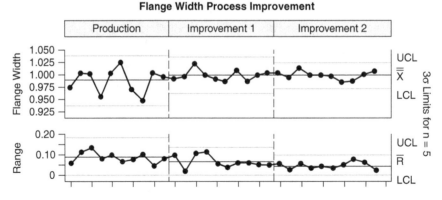

Figure 18.2 Control Chart Showing Process Improvement

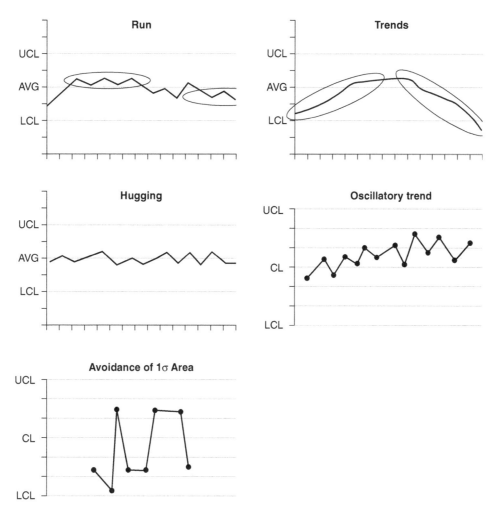

Figure 18.3 Examples of a Process Out of Control

sequential pattern that is not predictable. Statistically, they should not occur that way. Statistically, unpredictable data points can indicate cycling (often found in chemical processes) and stratification (hugging of points near the centerline). These problem conditions can exist even if all the X values fall inside the upper and lower control limits. In other words, an operator can fool himself into thinking nothing is wrong when something is indeed wrong.

As a general rule, a control chart is expected to look something like a badly mowed lawn with the grass at varying heights instead of all the same height. If a control chart has a pattern or trend in several data points, then something is wrong with the process. Randomness means there should be no pattern or trend. Reread the italicized words in the previous paragraph because they explain the following rules. These are rules for the interpretation of sequential data points on control charts. When one of the following patterns occurs in data from a process that usually produces normally distributed output, the population has changed and the process is no longer in statistical control (see **Figure 18.3** for a visual example of several of these situations):

    A. Nine points in a row on one side of the centerline
    B. Six points in a row steadily increasing or decreasing

    C. Fourteen points in a row alternating up or down

    D. Two of three points outside of the two sigma limits on one side of the centerline

    E. Four of five points outside of the one sigma limit on one side of the centerline

    F. Fifteen points in a row inside of the one sigma limit

    G. Eight points in a row on both sides of the centerline with none in the one sigma limit

Statistically, the probability of any one of these circumstances occurring is only about 3 out of a 1,000. When examining control charts, these patterns that indicate out-of-control processes may not initially be easy to see, but they become noticeable with practice.

SUMMARY

Process operators monitor control charts and determine what they are indicating about the process. Companies in the process industry use statistical thinking and statistics to help them determine when to make changes to a process. Changes are not made unless proof exists to substantiate the change. Statistical process control (SPC) is a tool used to continuously monitor process performance with charts and graphs. A company with a production process must use SPC to define nonconformances (products out of specification), assist in eliminating them, and to define the process capability.

The purpose of a control chart is to:

- Show the external customer the company's reliability as a supplier.
- Improve the process for the internal customer by removing fluctuations and disturbances.

SPC charts and associated problem-solving techniques provide a picture of the performance of a process. This picture can be analyzed to detect an incipient problem and make a correction before the process produces off-specification product. Further, the chart can be analyzed to help identify the root cause of a problem—the first step toward problem elimination and prevention.

REVIEW QUESTIONS

1. Explain the usefulness of SPC charts.

2. List five advantages of SPC charts.

3. Define the following: *attribute data, variable data, central tendency, range, standard deviation*

4. Explain what information an $\overline{X}$ chart reveals that an R chart does not.

5. Explain what information an R chart reveals that an $\overline{X}$ chart does not.

6. List the rules for determining whether a process is out of control, even if all data points are within the $\pm 3\sigma$ limits.

GROUP ACTIVITIES

Divide into small groups and do the following activities. Be prepared to discuss them or turn them in as an assignment.

1. This is a practical exercise on constructing $\overline{X}$ and R charts using the data in **Table 18.4.**

    The finishing tower on the styrene unit has had eight trays added to it to increase the purity of the finished product. The tower height increased by sixteen feet. Design engineers predicted the tower overhead temperature should average 308°F. The overhead temperature is critical (on average) for controlling the upper one-third of the tower temperature. It will control the amount of reflux returning to the overhead. A value this critical must be monitored, tracked, and

Table 18.4   Tower Overhead Temperatures

Day	Temp °F 8:00 A.M.	Temp °F 4:00 P.M.	Temp °F 12:00 P.M.
1	300	312	306
2	301	304	311
3	304	306	307
4	308	309	309
5	306	305	308
6	309	313	305
7	309	308	307
8	307	302	315
9	308	312	306
10	305	306	303
11	312	310	309
12	308	311	304
13	311	310	309
14	304	311	303
15	306	313	308
16	306	312	311
17	309	307	309
18	310	304	307
19	314	308	312
20	311	310	311
21	309	306	309
22	308	309	310

recorded. It also should have a control chart. The unit engineer assigned a crew to gather data to be used to construct the control chart. The unit crews took a thermocouple reading at the top of the tower every eight hours (some of the crew lost weight climbing the eighty-foot tower), and recorded it on a check sheet. After twenty-two subsets of data were collected, the engineer asked the crew to create a control chart on paper for him to look at. He would verify the data and calculations, and if they were okay, he would input the data into the unit engineering station and create a control chart online. The data collected are shown in Table 18.4.

a. Write a short paragraph telling the engineer what the control chart reveals about the tower overhead temperature population.

b. Use the data in Table 18.4 to construct an SPC chart for the engineer. Remember, this is part of the job for continuous improvement. It is not just the engineer's or management's unit; it is the crew's unit also. For the unit to remain in production, it must show a profit. The SPC chart created is one way to maintain on-specification product and to prevent rework and wasted resources.

2. Read and discuss the following article and answer the question at the end of the article.

A Control Chart Story

By Mike Speegle, February 2008

What can a control chart do for a process? Let's take an example from a compact disc coating operation at a plant in Dallas in 1983. Each hard memory disc is 100 percent electronically inspected for its functional performance. The process had been operating at a level of 60 percent to 70 percent yield (good product) for several years, slightly above the industry average at that time.

One of the critical factors affecting that performance is the thickness of the magnetic coating. Technicians in the coating room take production samples and measure the thickness on test equipment. If they feel it is too thick or too thin, they adjust the coating machine accordingly. They end up making frequent adjustments. This kind of "tampering with the system," as Dr. Deming calls it, actually makes things worse by adding more variation. Rather than tamper with the system, three things should have been done.

- First, let the process run without adjustment to determine the process capability.
- Second, make repeat measurements of the same piece of equipment to see how much the test equipment varies.
- Third, establish control charts for the coating thickness and train the operators to make the measurements and keep the charts updated. Operators were to have the authority to stop the process when a point showed an out-of-control condition.

To its dismay, that corporation learned that the test equipment had as much variation in its performance as the coating machines themselves. Next, it discovered that the coating process was much more stable than thought. It could run for a long time and perform within the thickness specifications without adjustment. The third and most surprising fact was that the operators began explaining all sorts of things that were wrong with the process, and they could relate them to variations on the charts. They had literally hundreds of suggestions, several of which were very valuable.

The corporation had always depended on its engineers to explain what was wrong, but now management was learning from the real experts, the people who were running the process. They saw things happen on a real-time basis and could link cause and effect. The coating process yield went from 68 percent to 94 percent in less than nine months. The same operators were running the process with the same supervision and help from the same engineers, but now everyone understood the information revealed by the variation, thanks largely to the control chart. The control chart was a key ingredient in changing attitudes and revealing the information that the process variation contained.

       a. List seven things the control chart revealed about this manufacturing process.

CHAPTER 19

Process Capability

Learning Objectives

After completing this chapter, you should be able to:

- *Explain what process capability reveals about a process.*

- *Explain what the numerical values 0.8, 1.0, and 2.0 reveal for process capability.*

- *Explain why process capability studies are conducted.*

- *List some problems solved by process capability studies.*

- *Explain what it means if the product specification range is the same as the common variation in the process.*

INTRODUCTION

Process capability or capability index (C_p) reveals whether a process (people, materials, machines, and methods) can meet product specifications and how effectively it can meet those specifications. Process capability measures the amount of statistical control in a process. It defines the relationship between the fluctuations inherent in a process and the specification or tolerance limits set by external customers. The higher the C_p, the smaller the risk of a measured value falling outside the specification.

THE SCIENTIFIC FOUNDATION OF A PROCESS CAPABILITY STUDY

With process capability, companies can make a connection between the statistical control they have set for the manufacturing process and the specifications of the product. For example, a customer may request tighter specifications on some contaminants in the styrene he is purchasing from the unit. The unit may not be able to meet that request because it is old and does not have the equipment or controls to meet those specifications. To try to meet that specification, the unit may produce significant off-specification product, which is economically unacceptable.

A process consists of people, materials (feed, catalyst, and so on), methods (procedures, SOPs), and machines. A good question to ask when doing a statistical study is: Is the process capable of doing what we want it to do? It is possible to calculate a value called the capability index (C_p), which gives clues about how capable a process is. The C_p may be thought of as a comparison of the engineering requirements to the statistical requirements. If put into an equation, it would look like:

$$C_p = \frac{\text{Engineering requirements}}{\text{Statistical requirements}}$$

Engineering requirements are usually stated in terms of upper specification limits (USL) and lower specifications limits (LSL).

A process capability study uses a formula to generate a numerical value that is directly related to the capability of a process to make specification product. The more that number is greater than 1.0, the greater the capability of a process to make on-specification product. On the other hand, if the value of the process capability is less than 1.0, the process won't be able to produce on-specification product consistently, if at all. Most customers expect a C_p greater than 1.5 units. This leaves a margin of safety for cases where the SPC efforts may not identify all assignable causes immediately.

A C_p of less than 1.0 indicates a generally incapable process. SPC will enable operators to do the best they can in the short term. However, in the long run, major process engineering changes will be necessary. Pressure on the operators to do the impossible—make on-specification product on a unit that is not designed for that specification—will only make matters worse by forcing operators to overcontrol the unit (tampering with the system), which will make the process even more unstable. A C_p of less than 1.0 is a long-term engineering problem that can only be corrected through redesign of the process unit.

Now, consider this a bit more. What is the main idea behind a process capability index? It is to determine that the process can consistently produce on-specification product (see **Figure 19.1**). A process capability study of a unit that has assignable causes interfering with the finished product quality would not be conducted. A process capability index is given only for a process in statistical control. Thus, all the assignable causes should be removed before attempting a process capability study.

The following are three components to process capability:

- The product specifications
- The centering of the natural process variation
- The range (spread of the variation)

See **Figure 19.2** for examples of how these three components affect process capability.

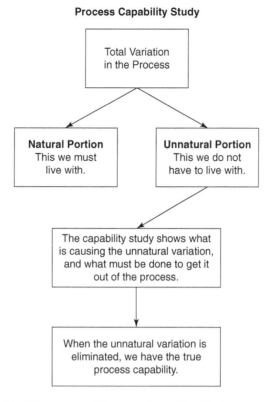

Figure 19.1   Graphical Summary of Process Capability Study

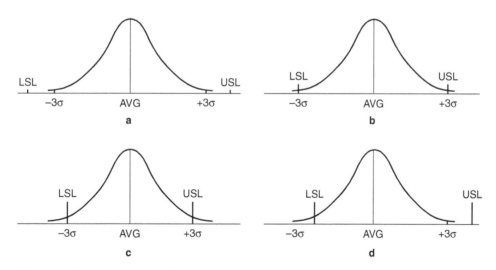

Figure 19.2   Product Specification and Process Variation

In Figure 19.2a, the upper and lower specification limits are far beyond ±3 standard deviations from the average. Obviously, the C_p is greater than 1.0, because the figure shows that more than 99.7 percent of the data is falling within the engineering requirements.

In Figure 19.2b—a case that almost never happens—the ±3 standard deviations fall exactly on the upper and lower specification limits. This process C_p would be exactly 1.0, because the process is doing what it should; however, it better be watched constantly, as a small shift in either direction would result in the production of off-specification product.

In Figure 19.2c, the specification limits are inside of ±3 standard deviations from the average. This means some off-specification material is resulting on both ends of the distribution, and drastic process improvements probably need to be made. The C_p of this process would be less than 1.0.

In Figure 19.2d, the process is "off center." The specifications are not being met on the low side. The C_p of this process would be less than 1.0.

If you are going to be a process operator, ask what the C_p is when you are hired and assigned to a process unit; you'll want it to be 1.5 and higher, so your workdays won't be very stressful.

Practical Examples of Specifications and Variation

It is tempting to conclude that a process should be modeled after Figure 19.2b, where the product specifications (engineering requirements) and common process variations are the same. However, remember that a normal distribution (±3σ) contains 99.73 percent of the output. For a process unit making 100,000 pounds of product a day, 270 pounds are off-specification product. Multiply this by 365 days, and you produce 98,550 pounds of off-specification product a year. If manufacturing TVs or automobiles, this means that 0.27 percent of output will fall outside of specification, which translates into 2.7 defects per thousand. This means a car with an average of 5,000 parts would contain 13.5 defects. This figure might be too small to seem critical, but if automobiles are subject to recall and repair, it could cost the automobile manufacturer millions. For something like the space shuttle with 1,000,000 parts, the defect rate of 0.27 percent becomes 2,700 defects, an unacceptable number for a vehicle in which the occupants cannot get out and walk away if something goes wrong.

Many companies have taken on the goal of design tolerances of six sigma (±6σ). This quality level translates into 3.4 defects per million. A six sigma quality level is the goal of world-class quality systems, but it cannot be achieved without closely analyzing process capabilities.

To put it in a nutshell, a process capability study refers to the inherent capability of a process to produce on-specification product under normal operating conditions. When the value of the process capability is larger, the inherent capability of a process to produce on-specification product is also greater. A higher value of process capability also means a lower level of product variation. And conversely, higher variability leads to a low process capability index.

From a practical viewpoint, a process technician may want to know which value of process capability is the best for his unit. Unfortunately, there is no simple answer to this question. As he attempts to achieve higher capability values, it will cost the company more in terms of modifying process design and control systems. As a rule of thumb, however, many plants regard a process capability value of 1.5 or greater as a practical compromise between high cost of design and poor quality. However, this number can change depending on the market conditions or change in the specifications of the product. If the design costs of the plants escalate, then it may not be economical to design processes with very high process capability indexes.

Many companies keep a running tally of process capability on their processes. Obviously, there will be variations in the capability numbers. For these companies, process capability is an indicator of process variability. They can look at the control charts or the process capability chart to get an idea about the day-to-day variations in the process capability, and hence, variability. Control charts can also be constructed for process capability indexes.

PROBLEMS SOLVED BY PROCESS CAPABILITY STUDIES

Many problems are essentially related to the nature and behavior of a distribution. They can be solved through process capability studies, because these studies provide a method for analyzing and changing distributions. They show the many different forms in which basically similar problems may present themselves. The following is a list of typical problems that can be solved through the proper application of process capability studies. Although the items on this list are not intended to be mutually exclusive, they indicate the broad scope of this technique:

1. Quality issues
    - Too many defects by operations
    - Product unstable or drifting
    - Wrong distribution (in case of distribution requirements)
    - Bad material entering the process
    - Too many adjustments (overcontrol)
    - Too much repair or rework
    - Too much scrap
    - Trouble in meeting schedules
2. New development/products/methods
3. Cost reduction
    - Automation
    - Machine and tool design
    - Purchase of new types of machines, test sets, and so on
    - Elimination of difficult or expensive operations

Many more categories and topics could be added to this list, but it was deliberately kept short.

SUMMARY

Process capability (C_p) reveals whether a process (people, materials, machines, and methods) can meet product specifications and how effectively it can meet those specifications. It measures the amount of statistical control in a process. It defines the relationship between the fluctuations inherent in a process and the specification or tolerance limits set by external customers. The higher the C_p, the smaller the risk of a measured value falling outside the specification. With process capability, companies can make a connection between the statistical control they have set for the unit and the specifications of the product.

A process capability study uses a formula to generate a numerical value that is directly related to the capability of a process to make on-specification product. The more that number is greater than 1.0, the greater the capability of a process to make on-specification product. A process capability index is given only for processes in statistical control.

REVIEW QUESTIONS

1. Explain what process capability reveals about a process.

2. Explain what the values 0.8, 1.0, and 2.0 reveal about a process capability.

3. (True or false) You would never do a process capability study of a unit that has assignable causes interfering with the finished product quality.

4. Explain why process capability studies are conducted.

5. List some problems solved by process capability studies.

6. Explain what it means if the product specification range is the same as the common variation in the process.

GROUP ACTIVITIES

Divide into small groups and do the activities below. Be prepared to discuss your conclusions with the rest of the class.

1. Management has just finished a process capability study for your unit. You find out the capability index is 0.8. Discuss what this means to you as an operator and what you can do to increase the numerical value.

2. Management has just finished a process capability study for your unit. You find out the capability index is 1.8. Discuss what this means to you as an operator and what you can do to increase the numerical value.

CHAPTER 20

Epilogue

INTRODUCTION

This book has covered a lot of ideas and concepts. You might not be able to remember all of them, but hopefully you will remember enough to understand how important quality is to a company and to a nation's economic survival. If you are in the workforce or are about to enter the workforce, you are, or will be, involved in a work process—making a product or delivering a service. Whereas you might have accepted a work process as it existed before reading this text, you should now be able to assess and improve processes. However, keep in mind the concepts of the next few pages.

THE ADVANTAGES OF QUALITY

The following summarize a few important concepts about quality:

- Quality is not a program; it is an approach to business aimed at performance excellence.
- The quest for quality never ends. It is an ongoing goal (marathon) until the company or individual ceases to exist.
- Quality is a collection of powerful tools and concepts that work and are applicable in every aspect of a business.
- Quality is achieved with people, not things.
- Quality is defined by external customer satisfaction.
- Quality includes continuous improvement and breakthrough events.
- Quality increases customer satisfaction, reduces cycle time and costs, and eliminates errors and rework.
- Quality is for all organizations that produce a product or service. This includes schools, health care, social services, and government agencies.
- Improved financial and performance results are the natural consequences of effective quality management.

Quality Is a Function of the Process

One of the most important facts mentioned in this book is that quality is a function of the process (system). If the process is flawed, the work will be flawed no matter how hard an employee works to prevent the flaw. Remember Deming's red bead experiment in Chapter 3? An employee is just a worker in a process. The process determines quality, and management designs the process. There used to be a myth that the worker was the problem and that all the problems would go away if the workers were all "quality" workers. Most companies today realize the fallacy of that myth. Management is responsible for 80 percent of quality problems. Employees making products and dealing with customers do indeed have the greatest influence on quality. However, they are only as good as the system in which they work. Most people want to do good jobs but, if a good worker is thrown into a bad system, the system will win every time. A "bad" system (one having a C_p less than 1.0) causes two things to happen:

1. Off-specification product is going to occur at some frequency, no matter what the operator does.

2. Morale will deteriorate, and employee buy-in will diminish. If the system is bad, has been bad for years, and management knows it is bad, why should employees exert themselves?

Management is responsible for building two types of quality into systems:

1. **Operational quality**—Processes must be driven by prevention and efficiently promote the production and delivery of high-quality products and services.

2. **Environmental quality**—Individuals should work in a setting that supports and enhances quality performance. Workers should be empowered and have decent working conditions and policies that support a quality improvement process.

Quality starts at the top. Besides the occasional pep talk, management must build systems—operational and environmental—that enable people to provide defect-free products and services to internal and external customers.

All Should Strive to Improve Quality

Quality should be organizationally—not departmentally (functionally)—focused. If each department (function) optimizes without regard to other departments, the organization actually suboptimizes and begins to develop a "silo" mentality, which is a restricted viewpoint. Workers become blind to the needs of other departments and seek only their goals. Although quality, productivity, and cost improvement potential exist within every department, the greatest opportunities lie in the interfaces between departments (in teamwork). A chef does not prepare whatever meal he pleases, hand it to a waiter, and instruct him to deliver it to whichever table the chef chooses. In the same manner, management should look at the way work flows and remove impediments between internal customer-supplier relationships to optimize organizational quality. Each department should strive to improve the quality of their interdepartmental processes.

Quality Is Created through Systems

Slogans, posters, and coffee mugs help build an awareness of quality, but that is all they do. The slogans and posters are not the systems that create quality. Many organizations have done excellent jobs at touting quality and educating people in some form of quality improvement process. This approach is motivational, but it will have little long-term impact without tools and systems in place. An organization's culture of quality is formulated and reinforced by management and its policies, and assisted through training and empowerment. Systems (processes), not slogans and posters, produce products and services. Management must provide the resources that create quality systems.

Statistical Tools are not Enough

SPC techniques can make significant contributions to organizational quality by serving as tools for identifying problems that contribute to quality fluctuations. Measurement is critical to any sustained quality effort, but SPC techniques and other statistical tools are only tools. They identify variation and problems and can help determine a method to eliminate the variation. However, without strategic, operational, and environmental systems that support quality, all the technical problem-solving tools in the world will not result in enduring and organization-wide quality.

Establish Performance Systems

Crosby, Juran, and Deming have shown that managers and management systems account for 80 percent of what are commonly called quality problems. Once management is on board the quality train and actively encouraging continuous improvement and prevention for all processes, it must establish an effective performance system that includes every individual in the organization. A performance system will ensure that:

- Quality expectations (standards) are established, and employees understand why they are important.
- Barriers to quality performance (defective equipment, lack of training) are removed.
- Regular, specific feedback on quality is provided.
- Rewards are given for quality performance.

With these environmental conditions, sufficient skills, and logical work systems, individuals can do what they prefer to do anyway—take pride in their work and produce high-quality products and services.

Empower Employees

Many process operators engage in activities outside of work that require expertise and mental agility. They may be city councilmen, run their own businesses, or serve on committees for churches and local governments. They may have hobbies that are challenging, such as astronomy, rebuilding automotive engines, or inventing, with several patents to their credit. Process operators should not be considered to be just a pair of hands that do only simple mechanical tasks. The survival of any company today is not dependent upon a small group of managers; rather, it is dependent upon the hundreds of employees that walk through the gate each morning. A failure to empower employees, even those who don't want to be empowered, can be a fatal mistake to a company. Remember, the scarcest resource in any organization is performing people.

Worker, Improve Yourself

Keep in mind that an organization cannot improve unless its people improve. Quality does not depend on equipment; it depends on the workers who run the equipment (process).

Continuous improvement—the Japanese call it **kaizen**—offers some of the best insurance for both the workers' careers and the organization's success. *Kaizen* is the relentless quest for ways to do things better and less expensively and to make a higher-quality product. Think of it as the daily pursuit of perfection. Every employee should practice self-improvement and keep stretching to outdo the previous day's performance, knowledge, and skills. Individual continuous improvements may come bit by bit, but enough of these small, incremental gains eventually result in valuable competitive employees. Because high-tech equipment with low-tech employees is a condition that leads to failure, it is best to remember that the best asset a company has is a highly trained and educated workforce.

REVIEW QUESTIONS

1. List four advantages to a company that has a quality system.

2. Explain what is meant by the sentence "Quality is a function of the process."

3. (True or false) A company with good statistical tools will have a good quality system.

4. (True or false) A company with good statistical tools and empowered employees does not need a performance system.

5. Explain your answer to question number four.

6. Explain why empowered workers are so important for continuous improvement.

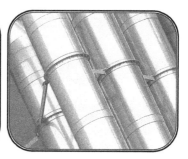

Glossary

80/20 rule–states that 80 percent of the problems come from 20 percent of the causes.

A

accuracy–the closeness of agreement between a test result and an accepted reference value.

advocates–customers who feel a strong connection to a company's product or service and will actively endorse it to others.

alignment–there are two types of alignment. *Horizontal alignment* extends from the external customer all the way through the organization to external suppliers. Horizontal alignment allows companies to determine customer needs and continually improve the ability to meet those needs. *Vertical alignment* extends from organizational goals and strategies to individual and team performance. It provides important information about how the organization chooses to spend resources to improve its overall performance and capabilities.

appraisal costs–those costs related to the detection of defects. They include the costs of inspection, testing, and other measures used to separate good products from bad.

assignable cause variation–(also called *special cause*) a factor that contributes to abnormal variation and is not normal to the process.

attribute data (also called *discrete data*)–qualitative data indicating the number of items conforming or failing to conform to specifications. Examples are good or bad product and percentage of late shipments to customers.

C

cause-and-effect diagram (also called *fishbone diagram*)–this diagram used to help people understand the complex relationship between an effect (a problem or goal) and its causes, and to aid in problem solving.

centerline–reveals the process average and is represented by the symbol $\overline{\overline{X}}$.

central limit theorem–a theorem that states that normal or nearly normal distributions tend to occur when working with averages.

central tendency–depicts where the values are centered (the *average* or *mean*).

characteristic–a property that helps to differentiate between items of a given sample or population.

chance variation–see *common cause*.

common cause–a factor that produces normal variation and that is expected in a process. Also called *normal variation* and *chance variation*.

competitive tool–any statistic, process, or concept that helps a company to improve work processes, reduce waste, and become more efficient.

conformance–the condition of being in agreement with specifications.

continuous data–see *variable data*.

control chart factor–a factor, usually varying with sample size, to convert specified statistics or parameters into a central line value or control limit appropriate to the control chart.

control chart method–the method of using control charts to determine whether of not processes are in a stable state.

control limits–limits on a control chart that are used as criteria for signaling the need for action or for judging whether a set of data does or does not indicate a state of statistical control.

controlled variation–variation due to chance causes and that is inherent in the process.

cookie-cutter teams–these teams are made up of people who think and act the same. They do not exhibit diversity.

cost of nonconformance–the expense of failing to conform to specifications.

cost of quality–the cost of prevention measures plus the cost of doing things wrong, of not conforming to requirements.

customer–(1) one who purchases a commodity or service. (2) Anyone—an individual or an organization—who receives and uses what an individual or an organization provides.

customer specifications—attributes or requirements requested by the customer.

D

data–facts, usually expressed in numbers, used in making decisions.

decentralization–when a major corporation decides to break itself up into three or four business units, or three or four smaller companies, in essence.

development–consists of the training received when people are hired, and all future training as business plans and equipment changes.

discrete data–see *attribute data*.

E

economics–the production, distribution, and consumption of goods and services.

economic resources–people, facilities, and raw materials.

employee empowerment–this means that management recognizes that employees can identify and solve many problems or obstacles to achieving organizational goals. Management provides employees with the tools and authority required to continuously improve their performance.

employee involvement–this means that employees and management recognize that each employee is involved in running the business and is necessary for the business's survival.

engineering concept of variation–has the object of meeting specifications that result in products that vary as much as possible because anything within specification is considered good enough. It does not seek to minimize variation or remove it.

entropy–the gradual decay and falling apart of things, similar to the aging process on the human body.

external customer–the customer who buys the finished product.

external failure costs–these provide measures of both product quality and customer satisfaction, and are costs that occur after delivery or shipment of a product or service to the customer.

F

failure costs–those costs associated with the correction of nonconforming material, including scrap, rework, repair, and warranty actions.

feedback–restatement of a message by a listener.

fishbone diagram–see *cause-and-effect diagram.*

G

globalization–the integration of national economies into the international economy through trade, foreign direct investment, capital flows, migration, and spread of technology.

H

hidden factory–a unit in a company where a group of workers is usually dedicated solely to rework.

I

infer–draw conclusions from presented or observed data.

innovation–new applied technologies that create a new dimension of performance.

input–the raw materials, services, equipment, training, manuals, facilities, and so on, required for a process to make a quality, finished product.

inspection–the process of measuring, examining, testing, gauging, or otherwise comparing the unit with the applicable requirements.

internal customer–the next person in a production system who adds value to the product or service to be sold to the external customer.

internal failure costs–those costs that are a measure of a company's operating efficiencies or lack of efficiency, including the costs of detecting errors after a product is made but not shipped.

ISO–an acronym for the International Organization for Standardization.

ISO 9000–the common name for a set of five standards sponsored by the International Organization for Standardization, that establishes requirements with which a company's quality management system must comply in order to export competitively to the European Community.

K

kaizen–the Japanese word for continuous improvement.

L

line of sight–workers' understanding of where they fit in and how they contribute to the chain of customers and suppliers in their organization.

lower control limit (LCL)–the farthest range below the centerline where a data point can be plotted and still be considered normal variation. The LCL is represented by -3σ.

M

metrology–the science of measurement.

monopoly–exclusive control, a commodity controlled by one party.

N

nonconformance–any deviation from requirements or specifications.

normal distribution–a fixed pattern of variation that can be described by a mathematical curve that is smooth and has a bell shape.

O

observation—(1) the process of obtaining information regarding the presence or absence of an attribute of a test specimen, or of making a reading on a characteristic or dimension of a test specimen; or (2) the attribute or measurement information obtained from the process. The term *observed value* is preferred for this second usage.

outputs–the finished product or service of a process.

P

performance standard–addresses schedule, cost, and specifications and how often we are required to meet them. For example, how often are you allowed to be late or absent from work before you get in trouble.

pie chart–a circular chart that reveals information in percentages.

population–all members or elements of a group.

precision–the closeness of agreement between test results obtained under prescribed conditions.

predict–create an argument from past to future using data.

prevention–planning or action that causes something not to happen.

preventive costs–those costs associated with activities designed to prevent defects. Such costs include participation in the design process to eliminate potential failure modes and process improvements designed to prevent production of nonconforming hardware.

process–any set of conditions working together to produce an outcome.

process capability–the limits within which a process operates based upon minimum variability as governed by the prevailing circumstances.

Q

quality–conformance to agreed-upon requirements.

quality assurance–the auditing of a process's quality control system.

quality control–the operational techniques and the activities that sustain a quality of product or service that will satisfy given needs; also the use of such techniques and activities.

quality control circle–a quality control circle consists of small groups of Japanese workers who meet voluntarily to discuss ways to improve their own work and to make suggestions to management for improving their manufacturing system.

quality system–a methodology that will cause or allow something desirable to happen or cause something undesirable to stop happening in a work process.

quality system standard–a standard that applies not to products or services, but to the process that creates them.

quality tools–a chart, graph, statistics, or a method of organizing or looking at things that help individuals and teams to continuously improve their work processes.

R

range–the spread between the values that are the farthest apart.

reactive costs–those costs that take effect only after off-specification goods or services have been produced or delivered.

registration accreditation board (RAB)–consists of groups of accredited registrars to whom firms apply to have their manufacturing processes registered.

representative sample–represents the population that exists at that time and place.

risk–the possibility of loss of value. When you invest in a stock and pay thirty dollars a share, you take a risk that the share may actually decrease below thirty dollars.

root cause–original reason(s) for nonconformances of requirements within a process.

S

sample–a small portion of a population that serves to provide information that may be used as a basis for making a decision concerning the population.

sample size–the number of units in a sample or the number of observations in a sample.

scientific approach–a systematic way to make decisions based on data rather than hunches; to look for root causes of problems rather than react to superficial symptoms, and to seek permanent solutions rather than quick fixes.

scope–from the first step of an activity to the final step of the activity.

sigma–the Greek name for the symbol σ, which represents the standard deviation.

special cause variation–an intermittent source of variation that is unpredictable. It is not part of the normal process. Also called *assignable cause*.

specification–a statement or enumeration of particulars required such as size, quantity, purity, delivery, etc.

spread—how spread out the values are, characterized as *range* or *standard deviation*.

standard–something established as a rule or a basis of comparison.

standard deviation–the measure of the dispersion of observed values from the mean (how far away data points are from the centerline).

standard of living–the minimum necessities or comforts held to be essential to maintaining a person or group in a customary or desired status.

statistic–a quantity calculated from a sample of observations, most often to form an estimate of some population parameter. To draw conclusions from numbers.

statistical control–a state of randomness.

statistical quality control (SQC)–the use of statistics to reduce to optimum levels nonrandom variation, and to assess and control acceptable random variation.

statistical process control (SPC)–the use of statistics to control a process.

subgroup–a set of units or quantity of material obtained by subdividing a larger group of units or quantity material, or a set of observations obtained by subdividing a larger group of observations.

surveillance audits–audits by registrars that review any changes to the quality system, ensure that corrective actions called for under previous audits have been carried out, and maintain the quality system.

system–(1) a group of equipment that works together to accomplish a task and/or (2) a collection of processes dependent upon one another to complete a task or product.

T

team–a group of people pooling their skills, talents, and knowledge to accomplish goals or tasks.

team dynamics–the interaction among team members.

tolerance limits (also called *specification limits*)–limits that define the conformance boundaries for an individual unit of a manufacturing or service operation.

total quality management–the totality of functions involved in the determination and achievement of quality for a business.

triology of management–an approach to cross-functional management that is composed of three managerial processes: quality planning, quality control and quality improvement.

U

uncontrolled variation–a pattern of variation that changes over time and can be attributed to an assignable *cause*.

upper control limit (UCL)–the farthest range above the centerline, where a data point can be plotted and still be considered normal variation. The UCL is represented by $+3\sigma$.

V

value-added work–work that really adds value from the point of view of the external customer for the product or service.

variation–the naturally occurring differences in things. Variation is the change or deviation in form, condition, appearance, extent, and so on, from a usual state, or from an assumed standard."

variable data (also called *continuous data*)–measurements that vary (such as temperature, pressure, flow) and may take any of a specified set of numerical values.

W

waste–anything that need not or should not have happened.

Z

zero defects–is about the understanding on the part of every member of the organization that processes must constantly be improved, and that defective systems must be reworked and reorganized from the top down. It is both an attitude and a performance standard.

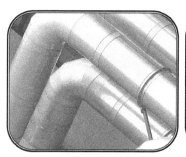

Bibliography

Alvin Community College, Department of Process Technology. "Quality, SPC and Economics of Process Technology." Alvin Texas, 1998.

Albert, Mike. "Five Steps to Stellar Customer Service," *Quality Digest,* August 2007, pp. 49–51.

Aguayo, Rafael. *Dr. Deming, the American Who Taught the Japanese About Quality.* New York: Carol Publishing Group, 1990.

Berk, Joseph and Berk, Susan. *Total Quality Management.* New York: Sterling Publishing Company, Inc., 1993.

Cochran, Craig. "Measuring Service Quality," *Quality Digest,* March 2008, pp. 30–33.

Crosby, Philip B. *Quality Is Free: The Art of Making Quality Certain.* New York: McGraw-Hill, 1979.

Crosby, Philip B. *Quality Without Tears.* New York: McGraw-Hill, 1984.

Deming, W. Edwards. *Out of the Crisis.* 2nd ed. Cambridge, MA: MIT Center for Advanced Engineering Study, 1986.

Dobyns, Lloyd and Crawford-Mason, Claire. *Quality or Else.* Boston: Houghton Mifflin Company, 1991.

Drucker, Peter. *Management.* New York: Harper and Row, 1974.

Drucker, Peter. *Management Challenges for the 21st Century.* New York: HarperBusiness, 1999.

"Sun, Sand and Scalpels," *The Economist,* March 10, 2007, p. 62.

The Economist, April 2007, pp. 67–69.

"A Special Report on Innovation," *The Economist,* October 2007, pp. 3–19.

"No Lakh of Daring," *The Economist,* January 12, 2008, p. 58.

Harrington, H. James. "Basic Block and Tackle," *Quality Digest,* March 2008, p. 14.

Harrington, H. James. "The Decline of U.S. Dominance—Part 2," *Quality Digest,* May 2008, p. 12.

Harrop, Froma. "Welcome to the High-Tech Age of Consumer Jihadists," *Houston Chronicle,* April 13, 2008.

Munro, Robert. "The Devolution of Quality." *Quality Digest,* April 2007, pp. 28–31.

Okes, Duke and Westcott, Russell T. *Certified Quality Manager Handbook.* 2nd ed. ASQ Quality Press, 2001, pp. 29–30.

Pall, Gabriel A. *The Process Centered Enterprise: The Power of Commitments.* Boca Raton, FL: St. Lucie Press, 2000.

Paton, Scott. "Quality After Thoughts," *Quality Digest,* November 2007, p. 64.

Robitaille, Denise. "Bringing Your QMS to the Boardroom," *Quality Digest,* November 2005, pp. 41–48.

Sashkin, Marshall and Kiser, Kenneth. *Putting Total Quality Management to Work.* San Francisco: Berrett-Koehler Publishers, 1993.

Western Electric Company. *Statistical Quality Control Handbook.* North Carolina: Delmar Printing Company, 1985.

Index

F

failure costs, cost of quality and, 167–169
feedback, as communication tool, 111
Feigenbaum, Armand, 7
fishbone diagrams, 195–196
flowchart symbols, as quality tool, 188–189
Ford, Henry, 19
Ford Motor Company, 74–75
forming stage of team development, 95–96
Fourteen Point quality management plan (Deming), 28–29
Fourteen Steps to quality improvement (Crosby), 33–34
freewheeling brainstorming, as quality tool, 180

G

General Motors, 3
gestures, as communication tool, 108, 112–114
global competition, economics of, 140–141
global consumption, economics of, 139–140
globalization, defined, 138
global markets
economics of, 138–139
quality control and, 21–22
goals and objectives
of economic systems, 132–133
of teams, 95
government regulation, of competition, 134–135
grand average, control charts, 233–236

H

Harley-Davidson, 221
hidden factory model, cost of quality and, 164–166
high-maintenance individual, 117
histograms, 196–198
statistical process control, 221–223
horizontal alignment, in quality systems, 150

human relations system, organizational effectiveness, 123–124
human resources, quality control and, 21–22
Huntington, Collis P., 14–15
Huxley, Aldous, 20

I

IBM, global competition and, 140
iceberg model, cost of poor quality, 163–166
idea building, team dynamics and, 103
indexes of customer satisfaction, 74
individualization, facilitation of, 88
Industrial Application of Statistical Quality Control, The, 15
industrial economics, 137–138
industrial production, quality control and, 18–20
Industrial Revolution, statistical process control and, 219–220
Information Age, statistical process control and, 219–220
information management, communication and, 115
innovation, economics of, 135–136
internal customer
in quality systems, 151–152
supplier chain to, 148–149
total quality management and, 36, 59–60
internal failure costs, cost of quality and, 167–169
International Standards Organization (ISO)
American embrace of, 45–46
application process, 46–47
benefits of, 45
certification timeline, 48
contractual registration, 47–48
history and evolution of, 43
ISO 2000 system, 49–50
ISO 9000 system, 41–42, 45–50, 55
ISO 14000 system, 50
purpose of, 43–44
quality management process registration, 47

CPSIA information can be obtained
at www.ICGtesting.com
Printed in the USA
BVHW060003190722
642442BV00005B/64